The Fallen Angel Model

Deeper into the Mysteries

Joe P. Provenzano, Ron D. Morgan,
Dan R. Provenzano

En Route Books and Media, LLC
St. Louis, MO

En Route Books and Media, LLC
5705 Rhodes Avenue
St. Louis, MO 63109

Cover credit: Dr. Sebastian Mahfood, OP.
Cover Art: St. Michael defeats the Devil by Eugène Delacroix (1861). Mural on the ceiling of the Chapel of the Holy Angels in the Roman Catholic Church of Saint-Sulpice, Paris
https://smarthistory.org/delacroix-sulpice/
https://commons.wikimedia.org/wiki/Category:Saint_Michael_and_the_Dragon_(Delacroix)

ISBN-13: 978-1-952464-86-7
Library of Congress Control Number: 2021940485

Acknowledgments

We have been discussing the ideas in this book with many people over a long period of time and could never mention all of them. However, the following individuals encouraged our effort and helped us organize and present our ideas more clearly: Jay Braun, Bob Chamberlain, Tom Fraschetti, Kevin Hinzman, Dr. Robert Hochberg, Sandy Morgan, Linda Provenzano, Gary Provenzano, and Ron Renckly.

Dr. Tom Sheahen, former Director of the Institute for Theological Encounter with Science and Technology (2008-2021) provided detailed reviews and suggestions for clearer presentations in many areas in the book.

Meet the Authors

The Philosopher — Joe P. Provenzano has an M.S. in Physics and is the author of *Conscious Energy* (previously published as *The Philosophy of Conscious Energy*) and *How to Believe in God and Science — In Three Easy Steps.*

The Deacon — Ron D. Morgan is a deacon in the Catholic Church and is a lifelong follower and student of physics.

The Scientist — Dan R. Provenzano has a Ph.D. in Applied Physics from the California Institute of Technology. Dan works with lasers and fiber optic sensors.

Joe and Ron are retired and live near Dallas, Texas. They have been brothers-in-law for over fifty years and have been developing the ideas and insights in this book for almost that long.

Dan is the son of Joe and his wife, Linda. He is currently an Optical Scientist working in Blacksburg, Virginia. Dan is married with three children and has been involved with these ideas all his life. He wrote the *ProWave Interpretation of Quantum Mechanics* when he was in graduate school in the late 1990's (see Appendix B).

A Note from the Authors

This book is intended for anyone who is seeking to reconcile their belief in God with the findings of modern science. Two of the biggest obstacles to belief in God are materialism and the problem of evil. The ideas in this book help overcome these obstacles.

In fact, that is exactly what happened with Joe and Dan. We have always been scientifically oriented and working through the logic of what we've done, and writing this book has helped our faith. We've become more comfortable with the reality of the spiritual, non-physical realm and have developed a logical solution for the problem of evil. Ron, who began this project with a strong faith, has seen it become even stronger by finding several connections between modern physics and theology. It is our hope that the ideas in this book will help others.

Table of Contents

Appendix A

Physics and the Not-So-Physical

Appendix B

Chapter 1

Mysteries from Science and Scripture

"The day science begins to study non-physical phenomena, it will make more progress in one decade than in all the previous centuries of its existence."

—Nikola Tesla

Science has given us amazing technology. Our day-to-day lives depend upon electronics and global communication. However, science for many can be a mystery: How does everything we use actually work? The first part of our book addresses one of the greatest mysteries in science as it relates to the underlying reality of physical substance. The scientific community anticipated that the fundamental particles that compose matter would behave as small, solid entities that obeyed Newton's classical laws of mechanics. It turns out that they did not conform to this expectation. Sometimes, they act like particles, and sometimes they act like waves with mysterious behaviors. Interpreting their underlying nature is not considered a matter of physics, and today most scientists simply use the mathematics of physics to make statistical predictions of the physical outcomes of experiments. Very few are even willing to discuss "what's really going on," i.e., the nature of the underlying reality.

Many people wonder about the origin of our planet, our solar system, our galaxy, and the other 100 or more billion galaxies. What did the very beginning of our early physical universe look like? Physicists propose that our universe began in a very small, very hot and unstable state. It then cooled, expanded, and evolved to the physical universe we see around us today. However, science itself cannot address how or why it began because there is no way to experimentally verify what, if anything, happens outside of spacetime.

Actually, many aspects of the early universe itself are mysteries. For example, scientists have discovered that the early universe had near, if not perfect, order. The famous physicist Michio Kaku has compared it to a perfect crystal that was somehow broken and then evolved. At this point, no one has any explanation for why our universe started out this way.

Perhaps the greatest mystery uncovered by modern science is something called the Fine-Tuning Argument. A large number of fundamental constants exist in nature that seem to be finely "tuned" to the values they are, and if any one of them is slightly different, then life and consciousness as we know them could not have evolved.

Scripture and theology provide spiritual wisdom revealing God's will for us in this life and a path to live with Him forever. Nevertheless, there are mysteries in Scripture and in the accepted interpretations of Scripture.

Scripture tells us that everything God created is good. However, it does not tell us exactly how God created the universe. Furthermore, we know that our world is full of suffering, death and evil. We wonder: Why would a good God create a world like this, or more simply, what is the source of evil or where did evil come from?

The most accepted explanation of the evolution of the human body that is consistent with religious belief is that God infuses an immortal soul into each human being. The mystery is how a non-physical entity (the soul) can affect a physical entity (the body)? Physicist and theologian Steven Barr acknowledged this mystery in his book *Modern Physics and Ancient Faith* (p. 226), by saying "... if there is something immaterial about the mind, how does it affect the brain and the body?"

There are also mysteries related to the fall of the angels. What are the general implications of their fall? In particular, how is their fall related to our universe: its creation, its evolution, and our spiritual battles with them?

These scientific and Scriptural/theological mysteries have been around for a very long time. There are no easy answers. There are currently two major paradigms about the underlying nature of reality. One is materialism, which says only the physical is real. The second sees reality as two separate realms: the physical and the spiritual. Both of these paradigms have been unable to provide new insights into these mysteries. In this book, we introduce two original ideas, which together lead to a new, third paradigm and deeper insights into these mysteries not otherwise possible.

We begin in the next chapter by reporting on what we know from modern physics experiments.

Chapter 2

Introducing the Not-So-Physical

"I am a Quantum Engineer,
but on Sundays I have Principles."
—John S. Bell

Not many people argue that the physical world is real. In fact, a lot of people believe that the only real, existing entities are physical entities. For these people, the words "physical" and "real" mean exactly the same thing. The common term for this position is materialism. There is no place for God, angels or non-physical human souls in materialism.

On the other hand, the vast majority of the people who believe in God agree that the physical is real, but they also believe in the existence of a non-physical reality. Most of these people believe that God created both the physical realm and the non-physical, spiritual realm as two separate realms.

Until about a hundred years ago, nobody imagined that particles could exist that were part physical and part non-physical. Then something really strange happened. Physicists discovered that the small, fundamental "particles" of nature exhibited some behaviors and attributes that they could not explain using physical terms. This was in addition to their other behaviors and attributes that were physical.

Nobody was happy with this. It was obviously unacceptable to the materialists. Even the believers could not make sense of fundamental particles that were part physical and part non-physical. Every possible effort was made to explain how these strange entities really were completely physical. Some of the greatest minds in the world tried and failed.

Then an even stranger thing happened. Just about everyone in the scientific community simply refused to think about or discuss the real existence of the non-physical aspects and non-physical behaviors of these entities. However, in this book we are simply going to accept what the physicists have experimentally discovered actually does occur in nature and follow this line of thought to its logical conclusion. We call it the "not-so-physical" and describe it in the next chapter.

Chapter 3

Describing the Not-So-Physical

"Indeed our sensory experience turns out to be a floating condensation on a swarm of the undefinable."

—Teilhard de Chardin

Around the turn of the 20th century physicists believed that classical physics, which describes all physical phenomena in purely physical terms, provided a complete description of the physical universe. Some even claimed that all there was left to do was measure things to the next significant figure.

In the late 1800's and early 1900's, that simple picture of physical reality started to change dramatically. Laboratory experiments began to show that light could behave as a wave in some instances and as a particle in other instances. Soon, it was discovered that all small particles displayed this strange, *not-so-physical,* wave-particle duality.

Even today, a hundred years later, no physical way exists to describe "what's really going on" during elementary particle movement. Appendix A contains several examples of behaviors that cannot be described in purely physical terms. The situation is even stranger in the very early universe where there was no matter, i.e., no mass, and even the laws of physics as we know them had not yet formed. The early universe is briefly described in Chapter 5 and discussed in more detail in Appendix A9.

However, before proceeding, we need to be more precise about what we mean by the terms: physical, not-so-physical, and non-physical.

By ***physical,*** we mean something that exists in reality that we can, at least in principle, detect and measure, and that only behaves in ways that can be described by the laws of physics.

The things we experience in everyday life provide simple examples of what we mean by physical entities, things like rocks, the air we breathe, and the moon.

By ***non-physical,*** we mean something that exists in reality that we cannot detect, we cannot measure, and that behaves in ways that cannot be described by the laws of physics. (For more detail on what we mean by "cannot be described by the laws of physics" see Appendix A5.)

There are a number of non-detectable "beables" that are required in the theories and equations of modern physics. Several examples that satisfy the non-physical criteria are provided in Appendix A3. Of course, angels and other purely spiritual entities are categorized as non-physical by these criteria.

By ***not-so-physical*** we mean something that exists in reality that has at least one physical aspect and at least one non-physical aspect.

For example, we call an entity not-so-physical if we can detect and measure it but find it has behaviors that cannot be described by the laws of physics. Fundamental particles such as the electron are examples of the not-so-physical. They are detectable and measurable but behave in ways that cannot be described by the laws of physics. For example, physics cannot describe the electron's behavior between the slits in the double-slit experiment and the point of impact on the detector. See Appendix A1 for more details. Another example of the not-so-physical is an entity called the vector potential field. This field is detectable but not measurable. It is not measurable because the actual value of this field at every point has no physical significance. See Appendix A2 for a description of the experiment that proved this unmeasurable field has real existence.

We said earlier that we were going to present two original new ideas in this book. Introducing these distinctions as new categories, or ways of looking at reality, is the first of the two new ideas.

Encountering non-physical aspects of physical reality is not a new phenomenon. In fact, it goes back to the very beginning of physics. Modern physical theories began with Sir Isaac Newton's theory of gravity. The problem of the non-physical has been with us since then. Here's a comment from Newton himself who realized that the force of gravity acts between material bodies that are separated from each other with nothing in between them. His references to the material and the immaterial mean exactly the same as what we mean by the physical and the non-physical.

> ...That gravity should be innate, inherent, and essential to matter, so that one body may act upon another at a distance through a vacuum, without the mediation of anything else, by and through which their action and force may be conveyed from one to another, is to me so great an absurdity that I believe no man who has in philosophical matters a competent faculty of thinking can ever fall into it. Gravity must be caused by an agent acting constantly according to certain laws; but whether this agent be material or immaterial, I have left open to the consideration of my readers" (Stanford Encyclopedia of Philosophy).
>
> — Isaac Newton, Fourth Letter to Bentley

Of course, the general theory of relativity has replaced Newton's theory and explains gravity is in terms of "spacetime curvature." However, it's still difficult to understand this in purely material terms.

In order to deal with the experimental discovery of these not-so-physical entities, physicists developed a theory known as quantum mechanics. It is not necessary to discuss the experiments in very much detail, or the theory at all, in the main part of this book. We have, however, provided several descriptions of experiments and a brief description of the theory in Appendix A. The chapters in Appendix A are provided as optional reading for readers who do not know about these areas, but are curious, and would like an introduction. Appendix B includes an original interpretation of quantum mechanics written in 1996 by one of the authors of this book. For a deeper discussion about the discovery, reality and meaning of the not-so-physical and more detail on what we mean by "cannot be described by the laws of physics" see Appendix A5.

The following table provides a summary of this new way of categorizing reality:

	Physical Existence	Real Existence
Physical Entities	*Yes*	*Yes*
Not-so-Physical Entities	*Partial*	*Yes*
Non-Physical Entities	*No*	*Yes*

Table 1. Kind of Entities vs. Types of Existence

There are important implications that come with accepting this way of categorizing reality. In other words, accepting the fact that the fundamental "particles" and certain fields discovered by physicists really do exist in nature as not-so-physical, and some likely exist as non-physical, has many significant implications. We will discuss four of the most important implications in the remaining pages of this chapter.

The first is that it renders the worldview of pure materialism, where physical equals real, as incomplete and unequipped to explain the blurry nature of reality. Since materialism has always been an obstacle to faith in God, clearly removing this obstacle based on modern physics makes it easier to believe in a spiritual reality.

The second implication is that it could provide an approach to help physicists with one of the toughest issues they are facing today. The issue is that there are new physics theories being developed that make predictions about the existence of entities that cannot be experimentally verified. These theories are not following the scientific method as currently understood because this method demands experimental validation. The problem is that the scientific method as currently understood is based on the assumption that the reality studied by physics is purely physical and, therefore, physically verifiable. However, modern physics experiments, as discussed in this chapter and in Appendix A, have shown that the reality studied by the physicists does not conform to the assumption that it is purely physical. This means that these new theories may be correct even if they make predictions that cannot be experimentally verified. Physicists cannot afford to have important new theories, that may be correct, discarded because of an outdated paradigm that limits reality to only what we can observe.

There is no easy answer to this issue. However, we think that our distinctions or categories about physical, not-so-physical, and non-physical provide the necessary starting point for addressing the question: What is the "scientific method" today?

The third implication of this view of reality is that it provides insights into the mysteries of complex numbers and why they are so important in modern physics in general and quantum mechanics in particular. The key point is that if we correlate the "real" axis with the physical and the "imaginary" axis with the non-physical, we gain insights into these mysteries. The discussion of this implication is provided in Appendix A10.

The fourth, and most important implication for this book, is that this way of viewing reality is needed to understand the Fallen Angel Model (FAM). This vision of reality, together with FAM, form a new paradigm which provides a deeper understanding of reality.

Before proceeding, we will make two points explicit. We have already presented the first one, but it's included here for completeness.

1. It is beneficial to view reality as composed of three categories: physical, not-so-physical, and non-physical.

2. Entities can be in one category, undergo some kind of change, and then be in another category.

A simple example of the second point is shown in the mysterious double-slit experiment. In this experiment, the "particle," for example an electron, exhibits a not-so-physical behavior. It does this while moving between the slits and the detector, but loses that behavior, at least for an instant, at the moment of detection. The detection is a physical process which occurs at a point. See Appendix A1 for more details. Another example is provided by the evolution of the physical universe itself. We will discuss how the universe began at the Big Bang in a very not-so-physical, or more likely non-physical, state where there was no matter, i.e., no mass. Then part of the universe evolved into not-so-physical particles with mass and some of that combined to produce physical entities. See Chapter 5 and Appendix A9 for more details.

We are now ready to take the next step which involves combining this view of reality with ideas from another discipline.

Chapter 4

Combining Disciplines

"In the history of science, ever since the famous trial of Galileo, it has repeatedly been claimed that scientific truth cannot be reconciled with the religious interpretation of the world. Although I am now convinced that scientific truth is unassailable in its own field, I have never found it possible to dismiss the content of religious thinking as simply part of an outmoded phase in the consciousness of mankind, a part we shall have to give up from now on. Thus, in the course of my life I have repeatedly been compelled to ponder on the relationship of these two regions of thought, for I have never been able to doubt the reality of that to which they point."

— *Werner Heisenberg, 1974*

Insights into the mystery of nature provided by the not-so-physical approach go beyond providing a more complete picture of reality and removing materialism as a roadblock to faith. They actually provide a way to link the findings of modern science with the basic beliefs of traditional faith in a way never done before. However, in order to do that, we need to combine some ideas from science with some ideas from Scripture. This approach will eventually lead us to a discussion of the Fallen Angel Model (FAM).

Physics and cosmology cannot explain what happened "before" the Big Bang because time as studied by physics began at the Big Bang. To consider the source and nature of the emergence of spacetime, we need to turn to another discipline and think about a different kind of "time." In the latter chapters, we need to use the phrase "before the Big Bang" and when we do so we are referring to a sequence of events in the spiritual realm. We are not referring to time as it is typically used, nor are we implying in any way that God is inside of time. These actions and events cannot be described by physical laws and physical structures. See Appendix A7 for a discussion of "spiritual time."

Before modern science was discovered and developed, theology was considered to be the final authority on matters of the spiritual and the physical. The "credibility pendulum" started completely on the theological side because there was no other side.

Eventually, modern science began to prove theology wrong in some predictions regarding the physical world. As a result, theology lost credibility. Unfortunately, theology's loss of credibility was not limited to the physical realm. Many scientists challenged the existence of God and everything but the purely physical. The credibility pendulum swung too far to the science side.

Then, all of a sudden, physics experimental results revealed that there is more to reality than purely physical objects. Actually, it's a lot worse than that. We discuss in the next chapter and in Appendix A9 how the physical realm, the physical constants, and even the laws of physics evolved from non-physical and not-so-physical fields. As we have pointed out, it is time to embrace this not-so-physical blur and recognize that our physical universe emerged from—and continues to consist of—not-so-physical entities for its very existence.

We live in one universe, but it has several realms or dimensions including the physical, the not-so-physical, and the non-physical realms. In order to include theology, we need to address the spiritual realm. By "spiritual" we are referring to God's Kingdom, which transcends the spacetime, physical realm, and includes angels, human souls, etc., as traditionally understood. The spiritual realm, which involves consciousness, is very different than the non-physical entities discussed in physics theories (See Appendix A3). Nevertheless, the spiritual realm certainly fits our criteria as being non-physical.

It's time to place the credibility pendulum in its correct place, giving proper credit to both science and theology. However, it is necessary to be careful, because past attempts resulted in both disciplines encroaching on each other's domains. We think the discovery of the not-so-physical is the key to success. Science needs to be focused on, and make predictions in, the physical realm. However, it is deeply involved with the not-so-physical and actually postulates the existence of non-physical fields inside the physical theories. Theology is necessarily focused on the spiritual realm but encounters spacetime as part of God's creation.

If one views the physical and the spiritual realms as two completely different realities, then combining these two disciplines will not be productive. History is full of failed attempts to combine them. However, if one views the physical and the spiritual as two different realms of one underlying reality, then combining science and theology has the potential to be productive. That's because we would have two different perspectives on the same underlying reality.

We have seen one side of this in the previous chapters where physics has led us to view the underlying reality of our universe as fields that are not-so-physical. St. Thomas Aquinas addressed the other side of this in his *Summa Theologica,* Volume I, Question 61, Third Article where he states that "...the angels are part of the universe: they do not constitute a universe of themselves; but both they and the corporeal natures unite in constituting one universe."

We agree with the findings of modern science and with St. Thomas, which see the universe, i.e., God's creation, as one reality. Therefore, combining physics and theology has the potential to give new insights.

Truth cannot change from one discipline to the next. Each discipline sees the same reality from a different perspective. In other words, there are some connections that cannot be made without combining science and theology. It is not enough to say that science and theology can agree. Rather it is much stronger; they must agree.

Theology tells us that God only creates good. The evil we see around us was not created by God and we will explore this mystery further after we introduce FAM.

The Bible was written long before the discovery of modern science and does not tell us exactly how God created everything. However, it does tell us that He did create everything and everything He created was good.

Genesis 1:1 "In the beginning when God created the heavens and the earth ..."

Genesis 1:4 "God saw that the light was good. God then separated the light from the darkness."

Genesis 1:31 "God looked at everything he had made and found it very good."

When one thinks about actually trying to combine theology with science in a real way, it becomes clear that the origin of the universe is a good place to start. Science begins at the moment of the Big Bang and studies everything that we can measure after that. On the other hand, spiritual events studied by theology happen outside of spacetime and outside of the realm of science. The origin of the universe is a unique point where these two realms meet. If we want to find some kind of cause-and-effect relationship between theology and science, then that would be a logical place to start. In the next chapter we will take a closer look at the origin of the universe.

Chapter 5

The Origin of the Universe

"The laws of science, as we know them at present, contain many fundamental numbers, like the size of the electric charge of the electron and the ratio of the masses of the proton and the electron .... The remarkable fact is that the values of these numbers seem to have been finely adjusted to make possible the development of life."

—Stephen Hawking

It is a mystery how our universe began or how it could have been created. We have two stories that help us understand creation. One is the story we read in the Bible where God created the universe in six days. The other story is from science where the universe begins in a Big Bang, expanding and evolving into the universe as we know it today over a period of about 14 billion years. Both stories have some problems. Currently, they do not fit together. Our Scriptures tell us that God created everything good, but there is evil in our world. Science tells us that what we observe now came about through evolution. Most believers are of the opinion that evolution can be reconciled with their doctrines.

As a result, today there are three basic views about how the universe came into being.

First, there is the purely materialist view that relies only on science and random evolution. There is no God in materialism; the universe began and evolved naturally. There are a number of problems with this view. One of the biggest problems is that science has found that the basic constants of nature seem to be "tuned" to be the very specific values they have. If they did not have these values, evolution as we know it could not have happened. In order to avoid some kind of tuning, which implies a designer, materialists postulate an infinite number of universes that have all the possible values of all the constants. Given that, at least one universe would have to have the right values for life and consciousness to evolve. We are obviously in such a universe, and it would seem to us to have been tuned to allow evolution as we know it. The remaining, infinite number of universes are not detectable. This solution is very unsatisfactory to us (the authors) and to a lot of people. See Appendix A8.

Furthermore, many scientists claim that science has determined that some of the advances in evolution are too unlikely to just have happened randomly because the universe is not old enough. So, even if an infinite number of universes does allow for one universe to have the right tuning of constants, many scientists argue that there is not enough time for random evolution to have made the progress it has made in that universe. For examples that there is evidence of design in our universe, see the work being done by the Discovery Institute. — https://www.discovery.org

Second, there is the purely literal interpretation of the Bible. The problem with this view is that it is inconsistent with what science tells us about the beginning and evolution of the universe.

Third, there is the view that God created the universe as stated in the Bible, but the details of the stories in the Bible cannot be taken literally. They necessarily had to be written in language that made sense to the people at that time. The purpose of the stories was to inform us that God created the universe, but not exactly how He did it. Today, we can discover the details of the process using scientific research. Anyone trying to reconcile science and faith needs to adopt some version of this view. The main problem with God's directly designing the evolving, physical universe is that we know that it is full of evil, suffering, and death. That would imply that God is responsible for all the evil, suffering, and death.

In this book, we will combine ideas from science and theology and propose a new paradigm offering deeper insights into these mysteries. However, before we can bring in ideas from theology, we need to first discuss what science tells us about the early universe. Appendix A9 provides quite a bit of detail on what science has discovered about the early universe. Here's an overview from that appendix that is sufficient for our purposes here:

> The Big Bang Theory is the most accepted cosmological theory about how our spacetime universe began and evolved. According to the Big Bang Theory, the universe began about 13.8 billion years ago as an expansion of space and time. It started as a collection of fields with no matter in a very small volume. These fields were highly ordered but very unstable and almost immediately started to expand and evolve. The expanding fields of the universe produced matter, stars, and eventually all the elements in the periodic table, and everything else we see in the universe today.

Given this discovery, it should not be surprising that fundamental particles studied by physics today, which are the building blocks of our physical universe, are not-so-physical. We are simply seeing the effects from the non-physical beginning of our universe.

We are now in a position to turn to Scripture and theology.

Chapter 6

Creation of the Angels

"...the angel's immateriality is the cause of why it is incorruptible by its own nature."

—St. Thomas Aquinas

The existence and activity of angels is an important aspect of Christian doctrine. They are discussed extensively in Scripture and theology. The angels are pure spiritual creatures, and they were created good. We believe that any discussion of God's creation must include the angels. Many early Church Fathers believed that the angels were created before the physical universe was created.

Saint Augustine said that in Genesis when God said: "Let there be light, and there was light," that this is referring to the creation of the angels. (St. Augustine, *The City of God*, Book XI, Chapter 9). He also said what is meant by: "God divided the light from the darkness; and God called the light Day, and the darkness He called Night." refers to the fall of the angels (Chapter 33).

Fr. Pascal P. Parente has also supported this position.

> With many of the Fathers of the Church we believe as very probable that the Angels were created long before the material world. They were certainly created before man, because we find them already distinguished as good Angels and fallen angels on man's first appearance on earth. (from Gen. 3:1ff.; 3:24) …
>
> And in St. Ambrose: "Even though the Angels, the Dominations and the Powers had a beginning, they were already there when this world was made." *Hexaemeron,* I, 5, 19. The same opinion is defended by St. Jerome *(Super Epist. ad Titum, I),* St. John Damascene *(De Fide Orthodoxa, III, 3),* and others. — *Beyond Space,* Rev. Pascal P. Parente, S.T.D., Ph.D., J.C.B., pgs. 10 and 11.

St. Thomas Aquinas addresses this question in his *Summa Theologica,* Volume I, Question 61, Third Article: "Whether the Angels Were Created before the Corporeal World?"

In this Article he states that "…the angels are part of the universe: they do not constitute a universe of themselves; but both they and the corporeal natures unite in constituting one universe." (This is the same quote as presented in Chapter 4).

This line of thinking causes him to think it's more probable that the angels and the physical universe were created at the same time. However, he does not take a firm position on this. He acknowledges that earlier Church Fathers believed the angels were created first, and that this position is "not to be deemed erroneous."

The belief that the angels were created before the physical universe is not contrary to Scripture or to Church doctrine. In fact, it is in agreement with teachings of many of the Church Fathers. Furthermore, as you will see in the next few chapters, our ideas fit very well with St. Thomas Aquinas' position that the angels and the physical universe constitute one universe.

Chapter 7

The Broken Vase

Author's Photo

Suppose you walk into a room and find some strange pieces of glass on the floor. You start to examine them and find that mysteriously many of them fit perfectly together to form nice, smooth, curved surfaces. Why do they do this? It seems like a total coincidence. All these jagged pieces are fitting together, yet they were not touching each other when you entered the room. You could think of several theories for why the pieces fit together:

1. Random: Maybe it's just luck that the pieces fit together.
2. Many Rooms: Maybe there are lots and lots of rooms with random pieces of glass, and given enough such rooms, there has to be at least one room where they fit together like this.
3. Perception: Maybe the fact that you think they fit together has something to do with how you are observing the pieces.
4. Necessity: Maybe there's something about this kind of glass that it has to be that way.
5. Designed: The pieces fit so perfectly together that clearly, they have to somehow be the result of a design.
6. Holistic Design. Maybe a vase fell and broke!

Let's explore the two cases where some sort of design is needed to explain what's going on: cases #5 and #6. These are two vastly different ways that design could be involved. #5 says that the designer designed each piece so that they would be curved the way they are and fit together the way they do.

Now we get to #6, where the parts were once part of a higher design, i.e., a vase. Of course, a vase designer should get more credit because a vase is much more useful and beautiful than the individual parts. Furthermore, the designer may not be responsible for the fact that it was broken. So, don't blame the designer and creator of the vase if you cut yourself while examining the pieces. Also, the creator may have created many more useful and beautiful items that did not fall, and they remain out of view.

Options 1-5, as you may realize, are depictions of reality put forth by prominent thinkers in modern times. One of the purposes of this book is to provide option #6 and introduce a new view of the origin of our universe. We do this by accepting the blurry nature of reality from modern physics and making the case that our universe is designed, not at the physical level, but at a much higher level. You will see later in this book how this simple concept can be used to provide insights deeper into the mysteries discussed in Chapter 1.

It's very interesting to note that physicists like Michio Kaku, quoted below, are starting to think about the beginning of the physical universe as a breaking of something with perfect, or near-perfect, order. They see our universe as being in a broken, or dare we say, in a "fallen" state.

> ...think of a beautiful crystal that shatters...but at the beginning of time when the universe was first created, that's when the crystal existed in its perfect form. We call it the Superforce. A single Superforce held this crystal together. But then we had the Big Bang which shattered this crystal giving us the shattered universe of today. When you look around you, and you see the different forces, moun-

> tains, clouds, planets — it's broken. We live in a horribly broken world, but at the instant of creation there was perfection... —Michio Kaku
> — https://www.youtube.com/watch?v=RUlVFzl_BJs

This quote comes just after the 4-minute mark in the video. Dr. Kaku also provides a similar discussion in his book: *Parallel Worlds: A Journey Through Creation, Higher Dimensions and the Future of the Cosmos* (p. 84).

Chapter 8

The Fallen Angel Model (FAM)

"It was pride that changed angels into devils; it is humility that makes men as angels."

—St. Augustine

Our theology and Scripture tell us that some of the angels rebelled and fell from heaven.

> Then war broke out in heaven; Michael and his angels battled against the dragon. The dragon and its angels fought back, but they did not prevail and there was no longer any place for them in heaven. (Catholic Bishops, *New American Bible*) — Revelation 12: 7-8

Luke 10:18: "Jesus said, 'I have observed Satan fall like lightning from the sky'." Many critical interpretations of this passage would favor the sentiment that Jesus was telling his disciples they would have power over the evil one. However, this is the exact language He used, as recorded in Scripture. The following quote is from Scott Hahn:

> Saint Augustine and Saint Ambrose insisted that the 'heavens' and the 'lights' we read about in Genesis represent the realm of pure spirits. (Physical light does not appear till several verses later, on the fourth day.) God created these angels of light, as He created everything to be 'good'. Yet He also created them to be free, because only free creatures can experience love. Love cannot be coerced, or it ceases to be love. So, God presented the angels with a decision, and some of them chose not to return His love. The book of Revelation seems to allude to this event, though in symbolic language, when it says that 'a third of the stars of heaven' (Revelation 12:4) were darkened (8:12) and cast down. — Scott Hahn, *Joy to the World: How Christ's Coming Changed Everything,* New York: Image, 2014 (pp. 84-85).

The idea that the fallen angels were somehow "darkened" implies that they lost something like power or energy when they fell. This idea has been with us for centuries. This darkening is reflected in many works of art, such as, for example, the image shown on the cover of this book.

St. Thomas Aquinas, who is commonly called the Angelic Doctor, also referred to them as "darkened." In fact, he had much to say about the fallen angels and what they lost when they sinned and were thrown out of heaven. In his great work the *Summa Theologica,* he presented his thoughts in a question-and-answer format. Each general topic or question is addressed by several articles in which there are a number of specific questions and answers. For example, Vol I, Question 64, "The Punishment of the Demons," the First Article is: Whether the Demons' Intellect is Darkened by Privation of the Knowledge of All Truth? Henceforth, we'll refer to articles by the short-hand notion provided in the *Summa*. In this case, it would be Vol I (Q. 64, A. 1).

The important thing for us to do now is to, as best as we can, consider what the angels lost when they were darkened and thrown out of heaven. In the following few paragraphs, we'll summarize what St. Thomas wrote about this in the *Summa*.

It's important to note that St. Thomas felt strongly that God did not take away anything from the fallen angels that was part of their nature.

> For it follows from the very nature of the angel, who, according to his nature, is an intellect or mind: since on account of the simplicity of his substance, nothing can be withdrawn from his nature, so as to punish him by subtracting from his natural powers, as a man is punished by being deprived of a hand or foot or something else. (Q. 64, A. 1)

He goes on to say in this article that although God did not take away any knowledge that came with their nature, He did take away some of the other knowledge that He had freely given them. That freely given knowledge was twofold: knowledge about "Divine secrets" and … knowledge which "produces love of God." The first kind was not totally removed but was reduced. The second kind was completely removed, resulting also in a complete loss of charity.

In Q. 62, A. 3 and Q. 109, A. 1, St. Thomas says that he believes that all the angels were created in sanctifying grace, which is not something that they would have had by their very nature. However, those that fell lost their sanctifying grace.

In Q. 62, A. 6, St. Thomas argues that more gifts of grace and glory were given to the angels who had more natural gifts.

In Q. 63, A. 7, St. Thomas argues that since the sin of the angels was a sin of pride, and not a propensity to sin, that the higher angels were more likely to sin.

In Q. 109, A. 4, St. Thomas argues that the angels are ordered by their nearness to God, and those nearer to God have power over those who are further from Him. Therefore, the good angels have power over the bad angels. The bad angels lost their angelic order and some of their power when they fell.

St. Thomas said that when the fallen angels lost these gifts, that they lost nothing that was theirs by virtue of their nature. In summary, St. Thomas described what they lost: some knowledge that was given to them, their love of God, sanctifying grace, their angelic order—or standing in relationship to God, and the power that came with that order.

From physics, we have learned that the universe began as unstable, non-physical and not-so-physical fields with high order, i.e., very low entropy, which naturally leads to the question: Why is this the case? Professor Robert M. Wald, University of Chicago, provides the following insight:

> It seems to me to be far more plausible that the answer to the above question as to why the very early universe was in a very low entropy state is that it came into existence in a very special state. Of course, this answer begs the question, since one would then want to know why it came into existence in a very special state, i.e., what principle or law governed its creation. I definitely do not have an answer to this question. But I believe that it will be more fruitful to seek an answer to this question than to attempt to pursue dynamical explanations.
>
> —https://arxiv.org/pdf/gr-qc/0507094.pdf

As we pointed out earlier, physicist Michio Kaku has compared the very early universe to a perfect crystal that was somehow broken and then evolved. At this point, no one has any explanation for why our universe started out this way.

Linking these events from science and theology, where one caused the other, opens the door to a new idea or model about the creation of the physical universe. We call this idea the **Fallen Angel Model, or simply FAM.** It is the second of the two original ideas we present in this book. Together, they provide a new, third paradigm for understanding reality.

The other two paradigms are (1) materialism and (2) assuming that the spiritual and the physical are separately created realities. This new paradigm offers insights deeper into the mysteries of science and Scripture that are not possible with either of the current two paradigms.

The Big Bang was the beginning of the physical, spacetime universe. It is a realm of impermanent order and rapid change. It's reasonable to consider that it came from the fall of the angels and that God saw it for what it was, an imperfect, evolving universe. He could have ignored it and focused only on the good angels remaining in heaven. However, that is not what He did. God saw the imperfect, evolving universe as good and full of potential, but needing a Redeemer. This kind of love is reflected in the parable where a good shepherd would go searching for one lost sheep. — Matthew 18:12–14 and Luke: 15:3–7

The FAM idea comes from combining events from physics and theology. We are presenting no new physics, but we are presenting a new speculative theological idea. Actually, what we are proposing is a possible cause and effect linkage of an event from science with an event from Scripture. This linkage puts together everything we have been discussing into a top-down model of creation. It is made possible because the not-so-physical forms a bridge between the non-physical and the physical, and entities can change from one category to another.

This model fits together all the pieces of the puzzle from both physics and theology in a comprehensive way that the other models do not.

Before we go any further, we need to make an important distinction between FAM and the idea that the physical universe is evil and came from evil sources. Perhaps the most common expression of this is Gnosticism. *The Catechism of the Catholic Church* (CCC) specifically expresses this in section 285:

> …Still others have affirmed the existence of two eternal principles, Good and Evil, Light and Darkness, locked, in permanent conflict (Dualism, Manichaeism). According to some of these conceptions, the world (at least the physical world) is evil, the product of a fall, and is thus to be rejected or left behind (Gnosticism)…

We realize that at first glance FAM sounds similar to Gnosticism, but it is very different. With Gnosticism, the substance of the universe itself is evil. With FAM, the substance of the universe resulted from what the fallen angels lost. They lost a number of God's gifts, especially order and power. These gifts were directly from God, were all good and in no way evil. None of the fallen angels' nature or evilness was in these gifts. However, there were consequences of their fall.

In order to better understand the idea behind FAM, consider the following analogy. When an electron goes from a higher to a lower energy state, a photon of definite wavelength and frequency is emitted. In the physical realm, we observe that light is emitted by the electron in changing states. Imagine the immense amount of spiritual energy that would have been released when the angels fell from their perfect order to their lower order, losing most of the gifts and power God had given them.

FAM describes the change of state resulting from the fall as the beginning of space and time in the Big Bang. We certainly cannot explain the exact mechanism of this change of state. This mystery is at the interface between the realms of theology and physics. Collaboration between physicists and theologians, focusing on the not-so-physical aspects of reality, might lead to some insights into this mechanism.

The fall of the angels and the resulting disorder is similar to hitting a drum too hard or picking a guitar string too hard. In both cases, the instruments break and can no longer function as before. The sin of the angels is a sin of pride or of wanting power beyond their nature. They wanted to be more than the creatures they were made to be. In this transformation, or change of state, spacetime began.

The fallen angels are spiritual beings that are connected to space and time.

> For our struggle is not with flesh and blood but with the principalities, with the powers, with the world rulers of this

present darkness, with the evil spirits in the heavens. - Ephesians 6:12

Current cosmology supports decreasing entropy looking back towards the Big Bang and even possibly zero entropy. Zero entropy is perfect order. The spiritual realm is perfect order and eternal.

We speculate that from the eternal order of heaven comes the impermanent order of space and time. This implies that the non-physical that comes out of the spiritual realm is "broken" and unstable. Some of it devolves into the not-so-physical, and finally, some devolves into the physical. This is consistent with what we have learned from physics and discussed earlier in the book.

Using the insights from this top-down model, we should be able see traces of the non-physical in spacetime matter and energy in the form of not-so-physical entities. Of course, this corresponds to what has happened as discussed in the earlier chapters and in Appendix A.

Up until now, the physical and the spiritual have been viewed as two separate realms, yet some scientists and some theologians have crossed over into in each other's disciplines. FAM offers a new paradigm for how the spiritual and physical are part of one reality and gives us a holistic view of theological and scientific inquiries.

We believe that truth is truth and is not different in each discipline but that each discipline sees a different aspect of that same underlying reality. That reality consists of the physical, the not-so-physical, and the non-physical. We just need to be careful to let physics do physics and theology do theology with no more and no less for each discipline.

The concept that the physical universe emerged from, and is part of, a larger, non-physical realm has been around for centuries. Hinduism, many other religions, including some Christian mystics, and modern thinkers have proposed different versions of this concept. Given the developments of modern physics, it's time to look at it again.

FAM is completely different from these earlier versions because it is consistent with both science and Scripture. Not only that, but FAM also provides insights into many of the mysteries in science and Scripture. Some of these insights are discussed in the next chapter.

Chapter 9

FAM and Deeper into the Mysteries

"Something deeply hidden had to be behind things."
—Albert Einstein

Scientific theories are judged by how well they can predict the outcomes of experiments. They are not judged on how intuitively pleasing they are in themselves. Quantum mechanics and relativity are perfect examples of this. They are both very counter intuitive, but they make correct predictions. On the other hand, paradigms, or worldviews about the basic nature of reality, cannot be tested by actual physical experiments. They can only be judged by their ability to provide insights into the mysteries about reality that we know from science and Scripture. Nevertheless, like scientific theories, they should not be discarded out-of-hand because they seem strange or unrealistic at first glance. In this chapter, we make the case that the paradigm, based on the physical, not-so-physical and non-physical categories combined with FAM, offers new and deeper insights into many of these mysteries.

In Chapter 1, we mentioned several mysteries that come from science and Scripture. We pointed out that the two common worldviews, or paradigms, are not giving new insights into these mysteries. Adherents in the first paradigm, materialism, claim that only the physical is real. Adherents in the second paradigm believe in the reality of the physical, but also believe in the reality of a separately-created, non-physical, spiritual realm.

In this book, we introduced two ideas that, when taken together, result in a new, third paradigm. This paradigm suggests that the reality of this universe consists of physical, not-so-physical, and non-physical entities. The not-so-physical entities have both physical and non-physical attributes or behaviors. This paradigm further suggests that our observable universe began because of non-physical (in this case spiritual) entities becoming unstable and evolving as described by science. This paradigm further suggests that this instability came from the fall of the angels which we learn about from Scripture.

Note that the current two paradigms have been around for centuries and there seems to be a well-established, but unstated, assumption that no other paradigms are possible. People are reluctant to discuss new paradigms. After all, what else could there be? The materialists say only the physical is real. Believers add a non-physical reality for the human soul and the spiritual realm. Doesn't that cover all the possibilities? No, it doesn't.

The not-so-physical category of physics and FAM suggests an addition to the existing paradigms. Accepting the reality of the not-so-physical, only means accepting what is demonstrated in the experimental results uncovered by modern physics. For over a hundred years we have been observing "particles" that show some physical and some non-physical attributes and behaviors.

Scientists and theologians do not have a way to discuss the underlying reality of these, part physical and part non-physical, particles. That's not surprising because these particles don't fit in either of the existing paradigms. Of course, it is always very difficult to consider changes to existing paradigms. However, the time has come to explore a new, third paradigm that includes the category of the not-so-physical to better understand the mystery of what physics theories and experiments have revealed.

In this chapter, we will summarize some of the key points made in the earlier chapters and show how adopting the not-so-physical FAM paradigm can give new insights that take us deeper into these mysteries. The first five mysteries discussed below come from science and the final three come from Scripture (and accepted interpretations of Scripture).

The basic nature of our physical universe — Modern physics experiments have demonstrated over and over that the fundamental "particles" of nature have both physical and non-physical attributes and behaviors. Obviously, materialism cannot explain this, but neither can the other existing paradigm. The idea of the not-so-physical is simply an acceptance that what the physicists have found is correct. Nature is what it is, not what we want or expect it to be.

The Complex Numbers and Quantum Mechanics — We mentioned in Chapter 3 that, if we correlate the "real" axis with the physical and the "imaginary" axis with the non-physical, we gain insights into the mysteries of complex numbers and why they are so useful in quantum mechanics. This discussion is provided in Appendix A10.

The nature of the very early universe — Scientists have developed an amazingly detailed description of the universe, but also have uncovered many mysteries. One of the most mysterious findings is that the very early universe had perfect or near-perfect order, much like a perfect crystal. Something "broken" or fallen from the spiritual realm, as described by FAM, is consistent with this initial high order.

The source of the very early universe — An unstable universe that evolves from a non-physical, or nearly non-physical, beginning is consistent with the idea that it came from an undetectable non-physical event. It would look to us as if it came from nothing. The fall of the angels could be the source of that non-physical event.

The fine-tuning issue — Scientists also tell us that many of the constants of nature have values that, if slightly different, would result in a universe that could not evolve life and consciousness as we know it. For most people, this is an obvious indication of design. However, as in the broken vase example, the design could be of something more holistic than simply the individual pieces. With FAM, there is a holistic perspective where the "pieces" come from a non-physical realm of life and consciousness. Therefore, it should not be a surprise that these pieces have something within them that allows a path back to life and consciousness.

The scientific method — We pointed out in Chapter 3 that scientists are now developing some theories that cannot be scientifically verified. The mystery is that this contradicts the very basis of the scientific method. Scientists are being forced to do this because the underlying reality itself is not-so-physical, perhaps even in some cases undetectable. We feel that the categories of physical, not-so-physical, and non-physical are the logical place to start addressing this increasingly important issue.

The fallen angels —-FAM provides insights into the fall that relate it to the creation and evolution of our universe.

The material body and the immaterial soul — The most accepted explanation of the evolution of the human body that is consistent with religious belief is that God infuses an undetectable, immortal soul into each human being. The mystery is how can a non-physical entity (the soul) affect a physical entity (the body). The current paradigm which completely separates the non-physical from the physical provides for no connection or interface. The concept of the not-so-physical can help with this mystery. With this view, one accepts as reality that the body is made-up of molecules and atoms, which are themselves made-up of not-so-physical, fundamental particles. Perhaps the non-physical soul can interact with the non-physical aspects of the fundamental particles in the body. This is consistent with everything we are presenting.

Where did evil come from? — We cover this mystery and the new insights that FAM has to offer in the next chapter.

Chapter 10

The Mystery of Where Evil Came From

The mystery of where evil came from, given a good God as creator, is one of the most difficult mysteries of all.

Scripture tells us that God created the universe and everything that He created was good. However, there is the mystery that the universe is full of suffering, death, and evils of all kinds. Grappling with the problem and source of evil is a serious pastoral issue as explained here:

> More people have abandoned their faith because of the problem of evil than for any other reason. It is certainly the greatest test of faith, the greatest temptation to unbelief. And it's not just an intellectual objection. We feel it. We live it. That's why the Book of Job is so arresting.
>
> The problem can be stated very simply: If God is so good, why is his world so bad? If an all-good, all-wise, all-loving, all-just, and all-powerful God is running the show, why does he seem to be doing such a miserable job of it? Why do bad things happen to good people?" — Peter Kreeft, The Catholic Education Resource Center
> —https://www.catholiceducation.org/en/culture/catholic-contributions/the-problem-of-evil.html

Atheists argue that the problem of evil proves there is no God.

> The problem of evil is the greatest emotional obstacle to belief in God. It just doesn't feel like God should let people suffer. If we were God, we think, we wouldn't allow it.
>
> The atheist philosopher J. L. Mackie maintained that belief in God was irrational, for if God were all-knowing (omniscient) he would know that there was evil in the world, if he were all-powerful (omnipotent) he could prevent it, and if he were all-good (omnibenevolent) then he would wish to prevent it. The fact that there is still evil in the world proves that God doesn't exist, or if he did, that he must be "impotent, ignorant, or wicked." — Matt Fradd, Catholic Answers
>
> —https://www.catholic.com/magazine/online-edition/the-problem-of-evil

An initial insight into this mystery of evil comes from St. Thomas Aquinas: "For evil is the absence of the good, which is natural and due to a thing." — St. Thomas Aquinas, *Summa Theologica,* Vol I (Q. 49 A 1). In other words, evil is not a "thing." It is a lack of something needed, or something broken, e.g., a broken leg. That is an example of a physical evil. Our world is full of physical evils like this which involve suffering and death.

Moral evil results from freely chosen conscious acts that are not in accord with God's will. Creatures with free will necessarily have the ability to choose evil or they are not really free.

> God created these angels of light, as He created everything to be 'good'. Yet He also created them to be free, because only free creatures can experience love. Love cannot be coerced, or it ceases to be love. — Scott Hahn, *Joy to the World: How Christ's Coming Changed Everything*, New York: Image, 2014 (p. 84; this is part of an earlier quote from Scott Hahn that was presented in Chapter 8.)
>
> Scripture speaks of a sin of these angels. This "fall" consists in the free choice of these created spirits, who radically and irrevocably *rejected* God and his reign. We find a reflection of that rebellion in the tempter's words to our first parents: "You will be like God." The devil "has sinned from the beginning"; he is "a liar and the father of lies." — *Catechism of the Catholic Church* (CCC 392).
>
> God created man a rational being, conferring on him the dignity of a person who can initiate and control his own actions. "God willed that man should be "left in the hand of his own counsel," so that he might of his own accord seek his Creator and freely attain his full and blessed perfection by cleaving to him" *Genesis* 17: *Sir* 15:14. "Man is rational and therefore like God; he is created with free will and is master over his acts. — St. Irenaeus, *Adv. haeres.* 4.4.3: PG 7/1, 983. — *Catechism of the Catholic Church* (CCC 1730).

The issue of physical evil is addressed in *The Catechism of the Catholic Church.*

> But why did God not create a world so perfect that no evil could exist in it? With infinite power God could always create something better. But with infinite wisdom and goodness God freely willed to create a world "in a state of journeying" towards its ultimate perfection. In God's plan this process of becoming involves the appearance of certain beings and the disappearance of others, the existence of the more perfect alongside the less perfect, both constructive and destructive forces of nature. With physical good there exists also physical evil as long as creation has not reached perfection. *Catechism of the Catholic Church* (*CCC* 310).

In the FAM paradigm, connections can be made that correlate physical evil with the free will action of the fallen angels. With FAM, everything God creates is good, namely the spiritual realm and the angels with free will. FAM provides the insight that the fall of the angels could have resulted in the beginning of the evolving universe. As pointed out above, one would expect an evolving universe to be a trial-and-error process full of suffering, death, and evil.

Appendix A

Physics and the Not-So-Physical

Appendix A1

The Electron

"The electron, as it leaves the atom, crystallizes out of Schrödinger's mist like a genie emerging from his bottle."
—Sir Arthur Stanley Eddington

Everything we touch and feel is made up of atoms. What we actually physically touch are the electrons and their fields. We do not touch the inside of an atom.

If waves pass through two narrow, parallel slits they will produce a particular interference pattern at a distance from the slits. This is true for all waves, whether they're light waves, water waves, or sound waves. When single electrons are sent through a double slit apparatus an interference pattern builds up as the screen records each individual electron impact. They show both wave and particle behaviors.

In the Davisson–Germer experiment (1923-1927), an electron beam was reflected off a surface of a crystal of nickel metal and displayed a diffraction pattern. In other words, it demonstrated that the electron beam behaved in a particle-wave manner.

There are two conceptually different ways for physicists to interpret the mystery of the wave-particle duality of small particles such as the electron. Both interpretations, however, ultimately require the electron to behave in ways that are not-so-physical because its behavior cannot be described before measurement. Both interpretations also involve an inherent randomness in the motion and the ability to predict trajectories is only probabilistic.

The most common of the particle interpretations is called the Copenhagen Interpretation. In this case, the electron is associated with a mathematical entity called the wave function. With this view, the "particle" is not-so-physical because it does not actually have a defined position or a defined velocity as it moves along.

Physicists use a mathematical equation of the wave function to calculate the probability of where the moving particle will be measured. At the time of the measurement, the mathematical wave function "collapses" and then it is certain where the electron encountered the measuring device.

The uncertainty principle of quantum mechanics applies to the electron. It states that the position and momentum of any quantum particle cannot simultaneously be known with complete accuracy. Also, unlike the photon, since it has mass, it can be at rest. But it never is exactly at rest. There is always a blur in exactly where it is and how much momentum it has. According to the quantum field theory interpretation, it is not really a particle, but an excitation in the electron field. The wave collapse is non-local because the collapse happens faster than the speed of light. Non-local is described in Appendix A4.

In the quantum world, the electron cannot be directly measured in a purely straightforward physical manner. The uncertainty principle must be taken into account for the specific meaning of a measurement. Quantum mechanics can only predict the probability of where the electron will be and the measurement itself is subject to the uncertainty principle.

The electron and all its configurations in all the various elements make up our contact with the everyday world. Atoms have specific shells of energy with a specific number of electrons. Atoms make up molecules, and molecules make up everything we call physical.

Yet the concept of the electron is blurry because physics cannot describe its behavior as a particle before measurement, and as a wave it has a non-local aspect in the wave collapse.

Appendix A2

The Aharonov-Bohm (AB) Experiment

The AB experiment demonstrated the real existence of a mysterious field, called the vector potential. It is a "real" field, i.e., effects of its presence can be detected in the laboratory. However, its actual strength cannot, in principle, be measured, and it behaves internally, according to Aharonov and others, in a non-local manner.

The vector potential and its effect in this experiment clearly illustrates the boundary of the physical and not-so-physical.

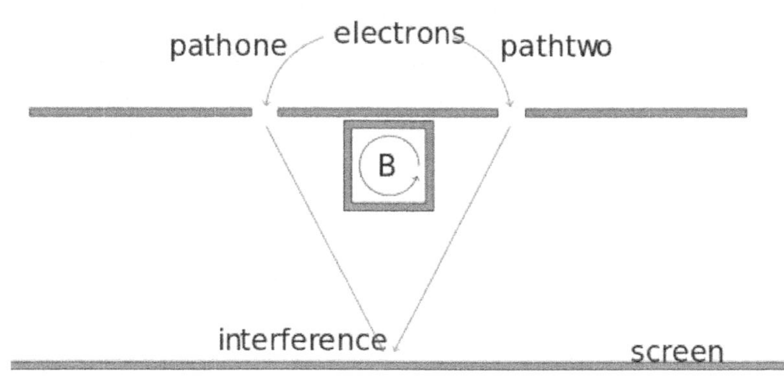

Figure A1. The Aharonov-Bohm Effect (a)

Courtesy of Sebastian Schmittner on Wikipedia
https://commons.wikimedia.org/wiki/File:Aharonov-Bohm_effect.svg

We explain this experiment in a very simplified manner. The solenoid in the illustration in Figure A1 is completely shielded, as indicated by the box in this figure, to ensure that there is no magnetic field outside of the solenoid. Then, the beam of electrons is split, and two electron beams are sent past the shielded solenoid. Then the two beams are recombined into one beam. When there is a magnetic flux present inside the solenoid the electrons exhibit a wave interference pattern on a detector screen as predicted by quantum mechanics. This wave interference pattern depends on the phase difference of the two electron wavefunctions.

Action at a distance in physics refers to one object affecting another object separated in space without actual physical contact between them. It appears that the magnetic field is acting at a distance. That it is somehow reaching out beyond the shield around the solenoid and affecting the beams.

To understand the AB effect, physicists attribute the interaction of a field called the *vector potential* with the electron wave functions. This field changes the phases of the electron wave functions to produce an interference pattern.

Before this experiment was actually performed, the vector potential was not considered a real field; rather, it was understood to be a mathematical property used to describe and simplify the calculations of the Maxwell equations describing electromagnetic theory.

In 1959, Yakir Aharonov and David Bohm proposed an experiment to demonstrate the potential terms mathematically present in the Schrödinger equations would do what we just described.

To explain the results of this experiment, it is necessary to choose between action at a distance from the magnetic field inside the shielded solenoid or conclude that the vector potential is a real field acting on the electron wavefunction. As a real field, it is detectable only in an indirect way. The vector potential is not a classical force field. The AB effect is a quantum effect.

We provide this quote from the famous physicist Richard Feynman.

> Feynman [13] has discussed the 'reality' (his quotation marks) of the vector potential. He gives two different definitions of reality (a) a real field is a mathematical function we use to avoid the idea of action at a distance. His second definition is (b) a real field is the set of numbers we specify in such a way that what happens at a point depends only on the numbers at that point. —A.M. Stewart arXiv2014v5 25/4/14 Page 7 of 12.
> —https://arxiv.org/ftp/arxiv/papers/1209/1209.2050.pdf

This is a quantum mechanical effect, and the apparatus has to be carefully set up to demonstrate the physics. Figure A2 is an illustration that shows the essence of the experiment.

The Aharonov Bohm effect describes a field that is real, detectable, but not measurable and has non-local properties. It is not measurable because its actual value has no physical significance. Only changes in the field over distance or time can be detected or measured. To study the non-local aspect, we provide a link to a paper by Yakir Aharonov, Eliahu Cohen, and Daniel Rohrlich.

— https://arxiv.org/abs/1502.05716

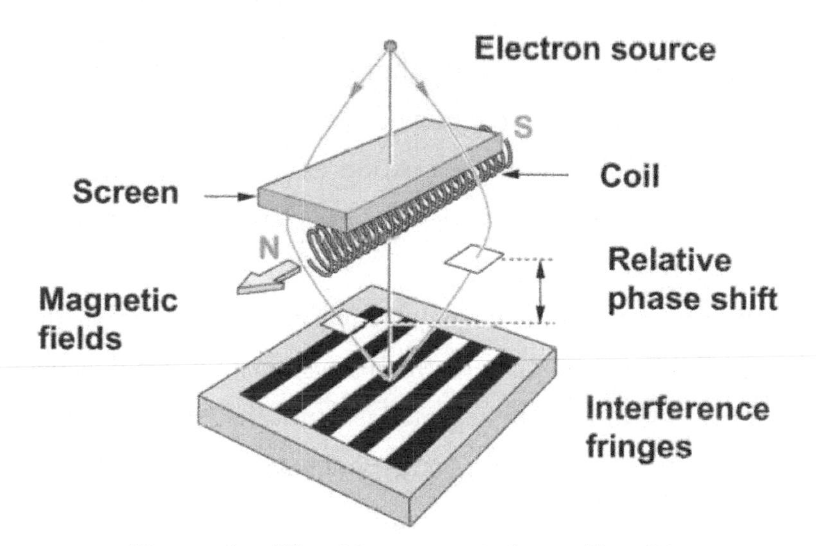

Figure A2. The Aharonov-Bohm Effect (b)

— https://www.ncbi.nlm.nih.gov/pmc/articles/PMC4323049/

The above link is to an excellent article that explains the AB effect research and results. These experiments illustrate a blur between the physical and the not-so-physical. The vector potential is real, but not-so-physical.

Appendix A3

Virtual Particles, Hypothetical Beables, and the Quantum Soup

When two particles collide, they can annihilate one another and create a new and different pair of particles. The mysterious "virtual particle" can be understood as mediating this interaction. It is not a "real" particle, but is transitory, short-lived, and undetectable.

We can think of the virtual particle as a short-lived disturbance in the underlying field. Such a disturbance is not as permanent as the fundamental particles themselves and only exists long enough to transfer the energy and momentum of an interaction.

Throughout this book, we have made a distinction between what is "real" and what is "physical." The virtual particle, the temporary disturbance, is real in the sense that it works in the calculations and can be understood as a disturbance of the underlying fields. However, since it's not detectable, it satisfies our definition of being not-so-physical, and perhaps even non-physical.

Physicists are currently exploring concepts that so far cannot be verified experimentally. They are interesting because they may be pointing towards future insights into the nature of physical existence. They definitely are pointing in the direction of the not-so-physical, and maybe the non-physical.

Sometimes, we run out of words trying to explain the entities of the quantum domain. We cannot escape the fact that the exact nature of the particle or the field cannot be pinned down. When we get to this point, it is difficult to stay in physics and to not drift into philosophy.

The term "beable" in theoretical physics can have many meanings. Just the very idea that it might be real gets it the label beable. It may exist for a very short time or under only very constrained conditions. It may be indistinguishable from other beables that are defined in a particular way.

Here are a few hypothetical beables/particles/fields:

a) Right after the Big Bang, cosmologists proposed there was an inflation period when the universe appears to expand very rapidly. We can see in the ultra-deep space photo taken by the Hubble telescope a mind-boggling density of galaxies. Through computer modeling, we can study the strands and fibers of the overall three-dimensional structure of the galaxies' locations. To help explain inflation, cosmologists consider the idea of an inflationary field. The quanta of this field are called inflatons.
b) A *tachyon* is a hypothetical particle that always moves faster than light.

c) The *axion* is proposed as a possible component of dark matter.
d) Weakly interacting massive particles *(WIMPs)* are hypothetical particles that are thought by many to constitute dark matter.
e) The *graviton* is a hypothetical elementary particle that mediates the force of gravitation in the framework of quantum field theory. The recent discovery of gravitational waves confirms the not-so-physical field nature of spacetime. Here are some references that detail the experiments:
https://brilliant.org/wiki/gravitational-waves/
https://www.scientificamerican.com/article/is-gravity-quantum/
https://www.quantamagazine.org/gravitational-waves-discovered-at-long-last-20160211/
f) A possible non-physical beable is the Goldstone Boson.

In the Standard Theory of particle physics, one of the four major forces of nature, the weak force, obtains its mass through an interaction with the Goldstone Boson. Bosons are one of the two fundamental classes of particles, the other being fermions.

The Goldstone boson is defined as a massless, scalar field with zero spin. From this definition it is easy to ask: "Is the Goldstone boson actually physical?

From an often-quoted Quantum Field Theory textbook:

"The fact that this mass is gauge-dependent is a signal that the Goldstone boson is a fictious field, which will not be produced in physical processes."

— *An Introduction to Quantum Field Theory,* Peskin and Schroeder (p. 734).

Many other presentations state that massless scalar particles do not belong to any reasonable theory of nature. We do not observe such particles. However, Goldstone bosons must appear in the theoretical equations.

We see them in the equations, but because they are non-physical, we only see the results of their presence in the laboratory.

g) Supersymmetry is a theory that proposes that every type of particle has one or more "super partners." If supersymmetry is true, every type of elementary particle in the Standard Theory would necessarily have partners that we have not yet discovered. There are ongoing experiments at CERN that are searching for physics theories beyond the Standard Model.
https://home.cern/science/physics/supersymmetry
https://atlas.cern/updates/physics-briefing/edge-susy

Of course, none of the particles, fields, or mixtures we have been discussing exist in isolation. We can think of the collection of a volume of all these beables and fields to be some kind of "soup." It is a big bowl of interactions and in its entirety a blended mixture of the not-so-physical and likely the non-physical.

> Isolated material particles are abstractions, their properties being definable and observable only through their interaction with other systems. —*Atomic Physics and the Description of Nature,* (1934) Neils Bohr

If we examine volumes of spacetime, some in matter-dense regions and others in near-vacuum regions, there will be self-field interactions and other field interactions. We will not be able to calculate the mixture of indistinguishable particles and all the possible kinds of superposition and entanglement of fields and their excitations. It would be an indescribable soup of not-so-physical existence. Julian Schwinger expresses the need for a holistic interpretation as follows:

> Is the purpose of theoretical physics to be no more than a cataloging of all the things that can happen when particles interact with each other and separate? Or is it to be an understanding at a deeper level in which there are things that are not directly observable (as the underlying quantized fields are) but in terms of which we shall have a more fundamental understanding? — *Quantum Mechanics - Symbolism of Atomic Measurements,* Julian Schwinger. https://www.goodreads.com/author/quotes/118871.Julian_Schwinger

Instead of trying to visualize one isolated quantum, we need to understand its existence in the context of its environment.

> This is all fiction, of course, due in part to the fact that bare particles are never seen in the physical world and can never be measured. — *Student Friendly Quantum Field Theory, Second Edition,* (2015) Robert D. Klauber (p. 318).

We discuss the early cosmology of the Big Bang in Appendix A9. The particle physics of the primordial soup of the unified fields at 10^{-43} seconds are being studied in the most powerful particle colliders in the world. See Figure A3 for a picture of The Large Hadron Collider (LHC).

Figure A3. A Picture of the Collision Section of The Large Hadron Collider (LHC) (CERN)

> Now the most powerful particle colliders in the world are recreating this primordial soup by heating matter beyond 3.6 trillion degrees Fahrenheit (2 trillion degrees Celsius). The hope is that a better understanding of quark-gluon plasmas can shed light on the evolution of the universe.
> —https://www.livescience.com/21715-big-bang-quark-gluon-plasma.html

We have reviewed some of the possible particles/beables/fields that can be incorporated into existing theories or used to develop new theories. The point we would like to make is that physicists need the not-so-physical beables to explain the blurry boundary that they would like to avoid. Of course, they would have preferred not talking about anything that is not physical, but they have to go where the physics takes them.

It is not wild speculation or some hidden agenda that brings us to this view. Physics itself is telling us that there are some not-so-physical, and perhaps non-physical, beables in their theories.

By including, rather than excluding, the other side of the physical boundary, we believe that it is possible to create new theories and insights that can lead to a deeper understanding of spacetime reality.

There is currently no complete theory which can keep track of the physical and the not-so-physical. It seems the more physicists press for clarity, the more they find "the blur."

Appendix A4

Non-locality and Entanglement

"The phenomenon of entanglement is the essential fact of quantum mechanics, the fact that makes it so different from classical physics. It brings into question our entire understanding about what is real in the physical world..."

—Leonard Susskind

What is meant in physics by non-local and entangled?

Imagine two particles separated by a large distance. What if we measure some value of one of the particles, which is undetermined until we measure it, and this allows us to instantly (or near instantly) know what the corresponding, previously undetermined, value of the other particle would be if we measured it? Since according to the laws of physics nothing can move faster than the speed of light, we would have a problem in figuring out the connection between them. However, quantum mechanics actually predicts that this will happen.

A faster than light interaction like this is referred to as a non-local effect. According to the laws of physics, two separate particles cannot interact like this. That would allow us to transfer information at a speed faster than light, and that is not possible according to current theories and never has been observed. However, according to quantum mechanics, "entangled" particles should be able to correlate quantum information between themselves this way. Unfortunately, however, we cannot use entangled particles to transmit information faster than light.

Entanglement is a quantum mechanical property that has to do with two particles interacting in such a way that they behave in many ways as a single particle. They are sort of mixed together and lose their individual properties.

An experiment to confirm the possibility of non-local effects between entangled particles was performed in 1982 by Alain Aspect. The experiment has been repeated with increasing accuracy many times since then. The results have always confirmed a non-local connection between the two entangled particles separated by a large distance. All attempts to explain this by saying that the variables actually have pre-existing values, but which are hidden from us, have been proven wrong.

At this point it is universally accepted that moving quantum particles do not have actual, specific values for position and momentum. There are also non-local effects that occur within an entangled pair of particles when a specific property of one these particles is measured.

In summary, the non-local effects of entangled particles that are predicted by quantum mechanics have been experimentally verified. These effects have been detected and measured many different times, but the non-local behavior itself cannot be completely described by physics. Entangled quantum particles are, therefore, not-so-physical according to the definition provided in Chapter 3.

Appendix A5

The Reality of the Not-So-Physical

As discussed in Chapter 3, the fundamental fields and particles of our universe, which have been discovered and are studied in physics, meet our description of being not-so-physical. We mentioned examples that include fields that are detectable, but not measurable, and the fundamental particles themselves, which behave in ways that cannot be completely described by the laws of physics.

By clearly describing the terms physical, not-so-physical and non-physical, we have established a distinction between "real" and "physical." There is a continuum among the physical, not-so-physical, and non-physical, but not between the real and the physical. Physical is real, but everything that is real need not be physical. That's one of the major points we are making in this book. The implications, three of which were addressed in Chapter 3, are very important. Making this distinction is only a natural and expected exercise because the world is having to come to terms with the blurry nature of reality.

There have been no new physics principles or laws developed or claimed here, but what is new is a labeling convention or description of the categories of not-so-physical and non-physical, yet both are real. We go into more detail here about what it means to say: "cannot be described by the laws of physics." It simply means that the complete behavior of a system cannot be explained with known laws of physics given enough information about the starting conditions. Thus, if the outcome is, at best, only statistically predictable, then any given single "particle" is not following any known law of physics.

An example of this is when an atom absorbs a photon. We can only discuss the "likeliness" of the absorption event, not that it will actually happen. The laws of physics for quantum mechanics deal with the probabilities and the statistical outcomes of events. Quantum mechanics is a very powerful tool, and even a macro-quantum system like a laser is very well understood by quantum physics. Yet down at the single quantum level, quantum mechanics can't predict which specific atoms will be involved in the lasing action with 100% certainty. Therefore, we call these quanta of matter and energy not-so-physical, but they are real. As stated earlier, a brief description of quantum mechanics is provided in Appendix A6.

Another example is the electron which is detectable and measurable. However, when passing through a double slit, there is no law of physics that can say which slit the electron will pass through or where it will land on the screen in the distance. However, the accumulation of particles is typically more important, and quantum mechanics is very good at giving this outcome. But again, down at the individual quantum level, it gives no certainty of any electron's individual path.

If it is a particle, then its path cannot be explained. If it is a spread-out wave, then its faster-than-light collapse to a point on the detector cannot be explained. Thus, even electrons can be said to be not-so-physical. However, electrons are real indeed.

These two examples can be contrasted with a baseball being hit with a baseball bat. As soon as the baseball leaves the bat, its entire trajectory can be described accurately with equations that predict which seat in the bleachers a home run hit will land. Another example of a predictable system is flipping a coin. It is commonly believed that if enough information is gathered about the initial velocity, rotation rate, air currents, floor roughness, etc., then it could be predicted whether a coin would land heads or tails. This is certainly NOT the case with quantum systems.

One consequence of accepting the reality of the not-so-physical is that it allows a reasonable, although necessarily not fully physical, explanation of the quantum experiments. A very short example is that the electron is a spread-out entity as it goes through both slits and then really does collapse in faster-than-light way at the detector. The full paper describing this interpretation is presented in Appendix B. It was originally developed in 1996 by one of the authors of this book after he realized that the electron has behaviors that are not-so-physical. It's called the ProWave Interpretation of Quantum Mechanics.

So, why hasn't it been necessary to make the distinction between "real" and "physical" until now? One of the main reasons for this is to keep the Standard Model of particle physics that the physics community in general has maintained for most of a century: Materialism is the law of the land. By blurring the line between real and physical, physicists are able to postulate all kinds of intermediate particles and force carriers but still shirk off any postulates of spirituality or non-physical reality because it can't be proven real. The discussions today have resolved not to discuss anything outside the current laws of physics such as the deepest meaning of the particle/wave duality. Confined to only dealing with macro-statistical behavior of quantum mechanical predictions, modern thinking puts no effort into a discussion of the non-physical or of spirituality. This "material world" worldview, which is referred to as materialism, has not only been challenged by our clear logic in this book, but has been proven wrong and is no longer a valid position to hold.

There are non-physical fields and excitations of such fields that exist in reality. They are undetectable, unmeasurable and behave in ways that are beyond the laws of physics. However, they are part of the very foundation of the physical part of the universe and essential to the formulation of the Standard Model of particle physics.

Appendix A6

Quantum Mechanics and the Not-So-Physical

"It is often stated that of all the theories proposed in this century, the silliest is quantum theory. In fact, some say that the only thing that quantum theory has going for it is that it is unquestionably correct." —Michio Kaku

There are many different ways to interpret the not-so-physical findings of modern physics. Almost everything that can be thought of has been tried, including trying to ignore the issue. However, the time of ignoring this issue has passed. In this appendix we'll look at a few of the more popular interpretations, and then suggest a path forward.

First, it will be very helpful to discuss the key postulates of quantum mechanics in a simple, most-accepted way.

- Every quantum mechanical system can be described by a mathematical "Schrödinger Wave Function" that evolves over time.

- The wave function depends on time, the positions of the entities in the system, and the forces acting on, and among, the entities. Note that the forces are actually embedded in the Schrödinger Equation in terms of potential energies.
- Only certain outcomes are possible, which are the specific matter wave solutions to the Schrödinger Equation. The outcomes are random for equal energy states, and the probabilities of these random outcomes can be computed by operating on the wave function at the time of the expected measurement.

The real postulates are more complicated than this, but this will suffice for our purposes.

Closely related to these postulates is something called the Heisenberg Uncertainty Principle, which states that you cannot know both the position and the velocity of a moving particle to arbitrary accuracy. In other words, the more you know about the position, the less you can know about the velocity (or momentum) and vice versa.

If you just use the postulates, and apply the appropriate mathematics, you can predict the statistical outcomes of experiments. Predictions using these postulates, and the associated mathematics, have been verified many, many times in all kinds of different experiments all over the world. The use of quantum mechanics is not controversial.

However, once the question: "What's going on before the measurement is made?" is asked, all kinds of disagreements and different interpretations spring up. Let's look at three of the many available interpretations.

We'll consider the Copenhagen Interpretation because it is one of the original and perhaps still the most accepted interpretation. According to this interpretation, the particle does not have a specific position or velocity before measurement, and, in fact, it does not make sense to even ask the question. However, at the moment a measurement is taken, the mathematical wave function randomly "collapses" to one of the possible outcomes.

The de Broglie-Bohm Interpretation argues that there is a pilot wave, which deterministically guides the particle in a nonlocal way. Key to this approach is the quantum potential term which can be derived from the Schrödinger equation.

Finally, we will consider the ProWave Interpretation, which is in line with quantum field theory, written by one of our co-authors. This original interpretation argues that the mass and energy of the "particle" are spread out in a wave packet before the measurement. At the instant of the measurement, the wave packet collapses in a faster-than-light, nonlocal event to the location where the measurement is made. The key difference is that there is never an actual particle, but rather the experiment demonstrates the wave nature of the electron. (See Appendix B: The ProWave Interpretation by Dan R. Provenzano)

There are dozens of other interpretations with different descriptions, however, there is no universally accepted interpretation among physicists.

The important point is that ALL of the interpretations about the most basic elements of reality using quantum mechanics and quantum field theory include aspects that are not-so-physical. This observation leads directly to a conclusion.

The underlying reality of our observable universe is not-so-physical. This means that materialism is no longer a valid description of reality.

In other words, the findings of modern physics and the fact that there is no purely physical way to explain the underlying reality described by modern physics tell us that the underlying reality itself is not purely physical.

Yet another way to say this is that our "physical universe" is not completely physical. It emerged from not-so-physical fields and remains dependent on them for its very existence.

Note that the wave function is a mathematical representation that corresponds to the underlying reality. However, the wave function itself should not be considered a real, existing entity.

This is consistent with quantum field theory (QFT), which describes the reality underlying our physical world as fields. These QFT fields, in themselves, are not detectable. We can only detect the excitations of these fields, which are the fundamental "particles" of nature.

Rather than say fields that have non-measurable attributes and exhibit non-local behaviors have no real existence, or cannot be discussed, we prefer to say that they do have a real, but not a purely physical, i.e., a not-so-physical, existence.

We are making a distinction between "real" and "physical." In other words, it makes sense to say that since these fields with not-so-physical attributes are needed for us to predict the way the physical world behaves, they must have some kind of real existence in spacetime.

For many people, "real existence" and "physical existence" mean the exactly the same thing, and this is a deeply entrenched concept. This concept made much more sense when the universe was thought to be fully explained by classical physics. However, that is exactly the concept that is disproven by modern physics. Consider the Table A1 below, first presented in Chapter 3, to help clarify this distinction.

	Physical Existence	Real Existence
Physical Entities	*Yes*	*Yes*
Not-so-Physical Entities	*Partial*	*Yes*
Non-Physical Entities	*No*	*Yes*

Table A1. Kind of Entities vs. Types of Existence

Physical, Not-So-Physical and Non-Physical Entities correspond to the descriptions provided in Chapter 3. Physical Existence means actually exists, outside of our minds, in reality, and is (at least in principle) detectable, measurable, and can be described by the laws of physics. Real existence means actually exists, outside of our minds, in reality, whether or not it is detectable, measurable, or can be described by the laws of physics. Everything that has physical existence has real existence, but not everything that has real existence has physical existence.

The physical existence and real existence of physical entities is obvious. Scientific results make a convincing case for the real existence of not-so-physical entities (fields and excitations of fields) based on the actual experimental findings of modern physics. See Appendices A1 and A2. Examples of potential non-physical entities, typically fields or excitations of fields, are provided in Appendix A3.

The distinction between physical existence and real existence is simply an admission that our ability to detect, measure and describe is not what defines reality. Not only is modern physics pointing us in the direction of the real existence of not-so-physical and even non-physical fields, but it's even giving us ways to visualize and imagine aspects of these realms. According to quantum field theory, fields are the underlying aspects of all types of reality.

In summary, the real existence of not-so-physical and non-physical fields is just an open acceptance of what physics is telling us and a willingness to see where it might lead us. This conclusion unlocks the door to new insights that have never been possible before. Some additional insights are provided in Appendix A10.

Appendix A7

Spiritual Time

Space and time as we know them were created at the moment of the Big Bang. Physics can start to explain the beginning of the universe at 10^{-43} seconds (Planck Time) after the Big Bang. The fields present at this time were unstable and their constant evolution is deeply connected with the passage of time.

Planck time is the time it would take a photon traveling at the speed of light to cross a distance equal to the Planck length. This is the 'quantum of time,' the smallest measurement of time that has any meaning in physics and is equal to 10^{-43} seconds. No smaller division of time has any meaning. Within the framework of the laws of physics as we understand them today, we can say that the universe came into existence when it had an age of 10^{-43} seconds.
— https://physlink.com/education/askexperts/ae281.cfm

As stated in Chapter 4, physicists cannot discuss a time "before" the Big Bang because time was created at the instant of the Big Bang.

The only way that we can conceive of God creating the universe is to think of God existing before and after He created it. These terms do not imply that God is somehow inside of time as we know it. He created time as we know it. So, when we say "before" the Big Bang, we are referring to God's time or some kind of "spiritual time." When we say "time," we are referring to time as we know it in physics, the regular time that's part of our spacetime universe.

There's not a lot we can say about spiritual time, but we can make some distinctions between the two kinds of time. We live in the physical world with its four known space-time dimensions of length, width, height (or depth), and time. God dwells in a different realm—the spiritual realm—beyond the natural perception of our physical senses. In the spiritual realm, the underlying structures are stable, but there is a sequence of conscious, intentional events. However, there is not necessarily any passage of spiritual time between these events. God is not limited by the physical laws and dimensions that govern our world (Isaiah 57:15). He is spirit (John 4:24).

God exists independently of time: "But there is one thing, my dear friends, that you must never forget: that with the Lord, a day is like a thousand years, and a thousand years are like a day" (2 Peter 3:8).

Appendix A8

Fine-Tuning Issue

"If everyone is thinking alike, then
somebody isn't thinking."
—George S. Patton

Science has found that a lot of "coincidences" related to the values of the fundamental constants of nature are needed in order for life and consciousness as we know them to have evolved. The coincidence topic was originally named the Anthropic Principle but is now more often called the Fine-Tuning Argument. The facts are not controversial, but the interpretations are very controversial.

Some of the facts that make up this issue are the following:

- The ratio of the mass of the proton to the mass of the electron must be very close to what it is for atoms to hold together;
- The charge of the proton and the electron are equal;
- The relative strengths of the nuclear and magnetic forces must be very close to what it is for atoms to hold together;
- The unique properties of water (ice floats);

-The expansion rate of the universe. If the expansion rate were much less than the actual value, the universe would have collapsed before life could have formed. If it were much greater, the formation of stars, which are needed for life, could not have formed;

- The list of coincidences goes on and on….

The point is that if any of these properties of nature were only slightly different—matter, life, and consciousness as we know them could not have evolved. Clearly, these coincidences seem to imply that our universe was somehow designed or tuned to produce life and consciousness.

Those who believe in God tend to be happy with this discovery and many have used it to argue that God must have directly designed the physical universe. However, if that is the case, then God is responsible for all the evil, suffering, and death in the universe. That problem is discussed in Chapter 10. Furthermore, as a method of producing life and consciousness, it is certainly not very efficient given the billions of galaxies in the universe.

Those who do not believe in God have had a lot of difficulty with this discovery. Their only argument is that there are an infinite number of separate universes. Each universe has a different combination for the values of the constants. Therefore, there has to be at least one universe that would have this seemingly "tuned" values for the constants. Since we have evolved, we must be in that universe. The other universes are not detectable. This is called the multiverse solution. Obviously, this is a philosophical solution, not a physics solution, because we can't test it experimentally.

William of Ockham was a philosopher who lived in the fourteenth century. He is most famous for a principle he developed called "Ockham's Razor." According to this principle the solution with the fewest assumptions is probably the correct solution. The multiverse solution has to be the greatest violation of Ockham's Razor in the history of philosophy. What could have more assumptions than needing to assume an infinite number of undetectable universes to make a position hold together?

The bottom line is that Fine-Tuning certainly shows essentially unarguable evidence of design in the physical universe. However, we believe that the design is at a higher, non-physical level. This new way of considering design at a higher level is part of the Fallen Angel Model (FAM) and is described in Chapter 8.

Appendix A9

Big Bang Cosmology

"Gravity explains the motions of the planets, but it cannot explain who sets the planets in motion."
—Isaac Newton

The Big Bang Theory is the most accepted cosmological theory about how our spacetime universe began and evolved. Nevertheless, it seems like the more we learn, the more we encounter new mysteries.

According to the Big Bang Theory, the universe began about 13.8 billion years ago as an expansion of space and time. Cosmologists believe that it started as a single "Superforce" in a very small volume. The four fundamental forces of nature: electromagnetic, gravity, the weak and strong forces were somehow contained in this Superforce. There were no elementary particles present. This Superforce was highly ordered but very unstable and almost immediately started to expand and evolve. The expanding fields of the universe produced matter, stars, and eventually all the elements in the periodic table, and everything else we see in the universe today.

To focus on the earliest instant of time, we refer to a time called the Planck duration of time, or Planck's epoch. It lasted only a tiny, tiny part of a second.

Note: This extremely short period of time that starts at the beginning and lasts only 10^{-43} second (which looks like this: 0.000000000000000000000000000000000000000001 second). Just to give you an idea of how short this duration of time is, there are a lot more Planck durations in one second than there have been seconds since the Big Bang.

The end of Planck's epoch is where cosmologists can begin to describe things. This is as far back towards the origin as cosmologists and physicists can consider. The period starting at this time is called the Grand Unification epoch and lasted only a very brief time. At the beginning of this time, the universe is thought to have been in a state of very low entropy. The special state of zero entropy represents no disorder. During this period, the four fundamental forces of nature, namely, the electromagnetic, weak, strong, and gravitational, were still united in a way that it would be impossible to individually distinguish them.

Inflation came next and very greatly expanded space in a tiny fraction of a second.

During these first three periods there were no particles. After inflation, electrons, quarks and the elementary particles started to form. The evolutionary process continued, eventually producing the universe we see around us today.

The photo in Figure A4, called the eXtreme Deep Field (XDF), was assembled by combining ten years of NASA Hubble Space Telescope photographs. They were taken in a very small patch of sky. Realize that we will see images like this in every direction we look. We are surrounded by an estimated one hundred billion galaxies. We encourage you to "fly through" the XDF at the NASA link, available online at https://youtu.be/odprMkzOst8.

Figure A4. The Hubble eXtreme Deep Field (XDF)
https://www.nasa.gov/mission_pages/hubble/science/xdf.html

As we observe the cosmos today, the galaxies are accelerating away from one another. Cosmologists still do not have a clear answer about the physical reason for this, but the most accepted explanation is something called "dark energy." It acts as a repulsive force, pushing the universe apart, causing it to pick up speed. Astronomers can only detect it indirectly, for instance, by measuring the distance between galaxies. There needs to be a lot of it. Dark energy is thought to comprise roughly 68% of the known matter/energy in the universe.

Similarly, physicists have had to postulate something else that is not detectable to account for the way the galaxies form and hold together. This time it's an attractive force like that produced by normal matter, but it's not visible. They call it "dark matter." Again, there needs to be a lot of it. The amount of dark matter needed is about 27% of the matter/energy in the universe. Combined together, this means that 95% of the matter/energy in the universe is unknown and can't be explained.

The important point from this appendix is that the fields in the early universe meet our definition of non-physical. They can't be described by the laws of physics because the laws themselves weren't even formed yet. It makes no sense to even talk about detecting or measuring these fields in a way that could be experimentally verified. Also, this early universe "Superforce" was at a very low entropy state. The story of the history of the universe as told by modern cosmology is a story of how the non-physical fields evolved into not-so-physical fields and finally into the physical universe we see around us today.

It's very important to note that, since we can't detect non-physical fields, it would look to us like the universe popped into existence from nothing. That's exactly the way it does look to us.

The amazing scientific discovery that our "physical" universe evolved from non-physical fields to the physical universe we see around us today deserves a lot more attention and thought than it has been given.

With the understanding that there was evolution from non-physical fields to not-so-physical fields to physical objects, we should expect to see traces of the not-so-physical remaining today. This is exactly what has happened. Finding traces of the not-so-physical is an understatement. Physics has actually discovered that the fundamental particles of nature, e.g., protons, electrons, and particles of light are, in fact, not-so-physical.

Appendix A10

The Mysteries of Complex Numbers and the Not-So-Physical

"The best that most of us can hope to achieve in physics is simply to misunderstand at a deeper level."
— Wolfgang Pauli

In this appendix we will show how correlating the categories of the physical, non-physical, and not-so-physical with the arithmetic of complex numbers provides insights into the mysteries of complex numbers. We will then present what others have said about the necessity of using complex numbers in quantum mechanics.

We also discuss how these three categories can help explain why complex numbers are necessary in quantum mechanics. We suggest that this representation is reflected in the complex phase of the quantum wavefunction and is an essential part of the foundations of quantum physics.

Let's begin with how complex numbers are defined.

$$c = a + bi$$

where a and b are real numbers. The number a is called the "real" part, b is called the "imaginary" part, and

$$i = \sqrt{-1}.$$

The first complete description of complex numbers with rules for their mathematical operations was published in 1572 by the Italian mathematician Rafel Bombelli. He did not use the word "imaginary." It is commonly believed that Rene Descartes first used the term "imaginary," and that he meant it in a derogatory manner. He said that we imagine these kinds of numbers, but they really don't exist. Of course, he had no way to know that they would become necessary in the development of the theories that describe the results of modern experiments. See *A Short History of Complex Numbers* by Orlando Merino for an article with more details.

—http://www.math.uri.edu/~merino/spring06/mth562/ShortHistoryComplexNumbers2006.pdf

Obviously, complex numbers are somewhat of a mystery in themselves. However, a bigger mystery is why they are not only so helpful in quantum mechanics but are actually needed in this theory. The following quote is from Professor Leonard Susskind: "The need for complex numbers is a general feature of quantum mechanics…." *Quantum Mechanics, The Theoretical Minimum,* Leonard Susskind and Art Friedman, Basic Books, 2014 (p. 44).

Some originally thought that what is really needed in quantum mechanics is just a pair of numbers. However, physicists quickly realized that in order to do the needed mathematical operations with a pair of numbers, rules are needed. The needed rules turn out to be the same as the rules for complex number operations. More discussion on this issue and some references are provided in the latter part of this appendix.

Let's now address the purpose of this appendix which is to show how adopting the concept, or paradigm, of the reality of the not-so-physical, along with the reality of the physical and the non-physical, can provide new insights into the mysteries related to complex numbers.

We'll start with the positive real numbers. The positive integers were most likely the first numbers used. Eventually, fractions, transcendentals, and the rest of the real numbers were being used. Somewhere in this process, we expanded the real number line to include negative real numbers. When we did this, we had to make sure that addition, subtraction, multiplication, and division between real numbers always produced real numbers. The issue of multiplication between two negative numbers needed to be either a positive or a negative number. Positive was chosen based on the logic that if a number is not, non-positive, then it is positive. The minus sign can be thought of as an operator that turns a positive real number into a negative real number and vice versa. The same solution was chosen for division.

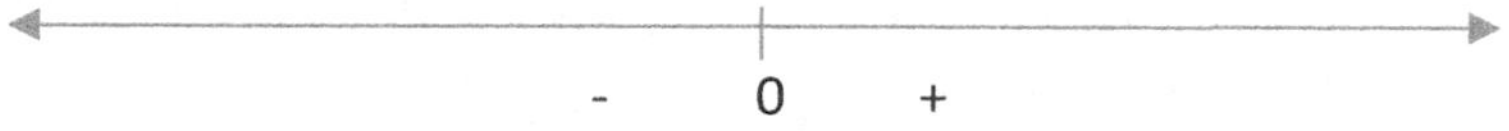

- 0 +

Figure A5. The real number line.

We have made the case that the strange results of the double-slit experiment and the other quantum results are telling us that our "physical" universe is not-so-physical after all. Within this paradigm, it follows that the "real" numbers, i.e., those that correspond with what we detect, measure and explain in physical terms could more accurately be called the "physical" numbers. These numbers are not able to describe what is going on when an electron is in a superposition of quantum states while moving between the double slits and the detector. Given this paradigm, it is reasonable that we would need a number system that allowed for states that contained non-physical and not-so-physical attributes and behaviors in addition to physical attributes and behaviors.

The complex numbers fit this situation perfectly. The imaginary axis is orthogonal (in a mathematical, not a space-like sense) to the real axis and complex numbers are part real and part imaginary. If we think of the real axis as corresponding to the physical aspect of reality and the imaginary axis as corresponding the non-physical aspect of reality, we see that a complex number can be interpreted as part physical and part non-physical.

$$c = a + bi$$

where a is the physical (P) part and b is the non-physical (NP) part. The number c fits our description of a real quantity that has both physical and non-physical aspects, and both really exist in nature. Note if $b = 0$, then c is a purely physical number, if $a = 0$, then c is a purely non-physical number. If both a and b are non-zero, then c is a not-so-physical (NSP) number. With the concept that the non-physical really does exist, maybe it's time to rethink using the word "imaginary" for this axis.

In quantum mechanics, equations with complex numbers are used to describe the system between measurements. However, at the point of the measurement, the equations produce probabilistic predictions of observables such as position, and these observables are always physical numbers.

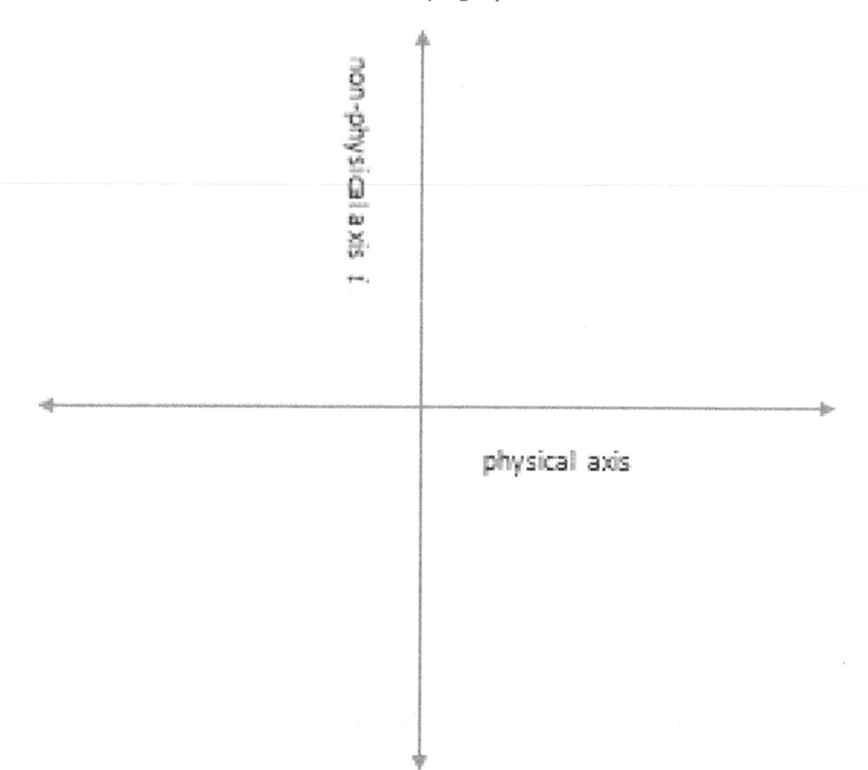

Figure A6. The complex plane with both axes considered to be real; one axis is physical and the other is non-physical.

We know that with complex numbers:

$$i = \sqrt{-1}.$$

But an interesting question is: Can we get some insights into why this is the case?

First, we know that i can't be a physical number because then $a + bi$ would be a physical number, and we would be back to physical numbers only. It has to be some kind of "non-physical" or "imaginary" number. In other words, it can't directly add, subtract, multiply or divide with physical numbers. It needs to be orthogonal to the physical numbers.

Let's apply the same criteria as we did for multiplication with negative real numbers. Let's require that the four standard processes in arithmetic on complex numbers must all produce numbers in the complex space. Addition and subtraction are easy; just add or subtract the physical and non-physical parts separately. The key is multiplication. Using "NSP" as a not-so-physical number we ask the question: What does i have to be in the equation below to keep NSP in the complex space?

$$NSP = (a + bi)(c + di)$$
$$NSP = ac + (ad + bc)i + bdi^2$$

The first two terms are in the complex space. The third term, which contains i^2, is like a minus times a minus, or think of it as not, non-physical. So, if it's not, non-physical, it is reasonable to think of it as physical. We're just using the same line of thinking as for the negative numbers. In other words, we can think that multiplying by i is an operator that turns a purely physical number into a purely non-physical number and vice-versa. So, let's make it physical and see what the implications are for i.

$$i^2 = k, \text{ where } k \text{ is physical,}$$
$$i = \sqrt{k}.$$

Note that k must be < 0 because if it's not, then i would be real, and we would collapse the space into purely physical numbers. Now let k have an absolute value of 1 because k is like a unit vector in the non-physical "direction." So,

$$i = \sqrt{-1}.$$

This also works for division. It can be easily shown that the result of the division of two complex numbers results in a number that is in the complex space.

For a complete and rigorous presentation of complex numbers, see the *Feynman Lectures on Physics, Volume 1: Algebra Ch. 22*

— https://www.feynmanlectures.caltech.edu/I_22.html

It might seem that just using a pair of numbers for the physical and non-physical would be sufficient for quantum mechanics. However, physicists quickly realized that in order to do the needed mathematical operations with a pair of numbers, rules are needed. The rules turn out to be the same as the rules for complex number operations as discussed in the two references immediately below. The first is an abstract from a physics paper by Goyal, Knuth, and Skilling:

> Complex numbers are an intrinsic part of the mathematical formalism of quantum theory, and are perhaps its most characteristic feature. In this paper, we show that the complex nature of the quantum formalism can be derived directly from the assumption that a *pair* of real numbers is associated with each sequence of measurement outcomes, with the probability of this sequence being a real-valued function of this number pair. By making use of elementary symmetry conditions, and without assuming that these real number pairs have any other algebraic structure, we show that these pairs must be manipulated according to the rules of *complex* arithmetic. We demonstrate that these complex numbers combine according to Feynman's sum and product rules, with the modulus-squared yielding the probability of a sequence of outcomes.
> — *Origin of Complex Quantum Amplitudes and Feynman's Rules,* by Philip Goyal, Kevin H. Knuth, and John Skilling. https://arxiv.org/pdf/0907.0909.pdf

Another example is provided in the following abstract from an article in the *American Journal of Physics*, written by Ricard Karam.

> Complex numbers are broadly used in physics, normally as a calculation tool that makes things easier due to Euler's formula. In the end, it is only the real component that has physical meaning or the two parts (real and imaginary) are treated separately as real quantities. However, the situation seems to be different in quantum mechanics, since the imaginary unit *i* appears explicitly in its fundamental equations. From a learning perspective, this can create some challenges to newcomers. In this article, four conceptually different justifications for the use/need of complex numbers in quantum mechanics are presented and some pedagogical implications are discussed.
> — https://aapt.scitation.org/doi/10.1119/10.0000258

One of the key places where complex numbers enter quantum theory is in the phase of a particle's wave function. As discussed in Appendix A6, wave functions are an integral part of quantum mechanics and are associated with each particle's movement. These wave functions have complex phases that are not observable. A very good description of this situation is given by Suzanne van Dam:

There are two ways to interpret the unobservable nature of the complex phases. In one interpretation it is concluded that states described by a different complex phase are physically the same and differ only in mathematical description. Another way to look at it is that the states are observationally indistinguishable, but are nevertheless physically different. *PhySci Archive Spontaneous Symmetry Breaking in the Higgs Mechanism,* Suzanne van Dam — http://philsci-archive.pitt.edu/9295/

Actually, there is third way, which is what we have proposed in this book. Limiting the interpretations to these two ways is a result of the traditional implicit assumption that the words "physical" and "real" mean exactly the same thing. By accepting that the complex phase actually exists in nature, but it is "not-so-physical" we have an insight into why the complex numbers are needed in quantum mechanics. With this approach, the complex, unobservable phase is considered to have a part that is non-physical.

Complex numbers are also used in the foundations of quantum theory. Here's an example:

1.1 First Postulate

At each instant the state of a physical system is represented by a **ket** $|\psi\rangle$ *in the space of states.*

Comments

• The space of states is a vector space. This postulate is already radical because it implies that the *superposition* of two states is again a state of the system. If $|\psi_1\rangle$ and $|\psi_2\rangle$ are possible states of a system, then so is

$$|\psi\rangle = a_1|\psi_1\rangle + a_2|\psi_2\rangle,$$

Where a_1 and a_2 are complex numbers. ...
— SUPPLEMENTARY NOTES ON DIRAC NOTATION, QUANTUM STATES, ETC. — R. L. Jaffe, 1996 — http://web.mit.edu/8.05/handouts/jaffe1.pdf

With the insight that complex numbers can be seen to represent or correlate to the non-physical and not-so-physical categories, one can see that these categories are present in the formulation of the quantum postulates.

In summary, considering that the fundamental particles of nature have physical and non-physical aspects provides a new insight into why the complex numbers are not only useful but are needed in quantum mechanics. All we have done is to consider the real axis as corresponding to the physical aspect of reality and the imaginary axis as corresponding to the non-physical aspect of reality. We have not suggested any actual changes to the physics or to the mathematics.

Appendix B

The ProWave Interpretation of Quantum Mechanics

Dan R. Provenzano

July 27, 1996

Abstract

It is widely accepted in Quantum Mechanics that measurements reveal the particle nature of elementary quanta, but there are many interpretations on how these "particles" move from the emitter to the point of measurement. This paper introduces in the ProWave (for "Propagating Wave") Interpretation of Quantum Mechanics. The basic idea is that elementary quanta always *exist* in the form of a wave, and always *travel* in the form of a wave, described by Schrödinger evolution, but are always *measured* each at a single location. This concept replaces all interpretations based on quanta traversing a particle path with the notion of a propagating wave coupled with a new concept of "Quantum Energy Localization." It is argued in this paper that the ProWave Interpretation explains all known experimental results in a "realistic" way that would have pleased Einstein, Schrödinger, deBroglie and all those who are currently looking for a sensible way to understand the implications of Quantum Theory. As examples, the 2-slit experiment, and EPR experiment, and a quantum eraser are interpreted in the ProWave picture.

Introduction

Classroom education on Quantum Mechanics (QM) concerns the various quantum phenomena and how to deal with them mathematically. In the laboratory, we use QM as a tool to predict measurement statistics. The experiments are repeatable, and the results are not in dispute. But when it comes to the interpretation of the theory of QM, there is much discomfort in the community. Sure, quantum phenomena are so removed from our daily experiences, and in many aspects are so counter intuitive, that it might seem only natural that our interpretation of the theory and corresponding view of reality must reflect that. Furthermore, the Complementarity Principle simply states that the elementary quanta are neither particles nor waves, but some entity that transcends both of their natures and only displays one of these attributes at a time. This principle is merely a statement of our conceptual difficulties with QM but doesn't really attempt a solution.

A semiclassical interpretation was put forth by Schrödinger who suggested that the wavefunction for matter waves is analogous to the field variables in electromagnetic waves. This interpretation was rejected long ago due to its intrinsic "nonlocality problems." However, the Bell Inequality [2] and its numerous experimental verifications have since dictated the need for a nonlocal interpretation of quantum theory. Thus, we can start with Schrödinger's original idea, and build on it to provide an understanding as to what happens during a measurement or "collapse of the wavefunction."

The ideas for the ProWave Interpretation resulted from desperately trying to make some sense out of the various experiments and connecting them with accepted quantum theory. We now think it is possible to (at least) conceive of how these bizarre quantum effects could come about in a sensible way.

The Experiments

To date, quantum theory has been overwhelmingly confirmed by physical experiments. The need for such a theory in the first place began as the experimental results of the early 1900's could not be explained by contemporary theories. For compactness, I will describe only the quantum interference effects of photons but note that electrons and neutrons have been observed to display quantum interference as well. Below is a review of three interesting experiments, our understanding of which is controversial. These provide a good survey of the weirdness of QM. In Sections 5 and 7, the ProWave Interpretation is applied to these experiments.

A) The infamous two slit experiment involves passing light through two slits and allowing the radiation from both slits to diffract and overlap on a distant screen. The interference fringes that are observed are attributed to the wavelike and coherent nature of the photons as they emerge from the slits. If the flux of photons is low enough such that one photon is present in the system at once, then each photon (event) produces only one localized detection at the screen. By integration of many such events, an interference pattern emerges, as predicted by the probability distribution of the wavefunction. Note that if the screen were placed sufficiently close to the slits so as to prevent overlap of diffracted waves from the slits, then each event induces local detection corresponding to the photon having been present at one of the two slits. In this case, the system probability is determined by an incoherent mixture of quantum states, where the quantum states refer to slit passage. The philosophical question raised by this whole experiment is: Does the photon pass through only one slit, or both, or can we even ask this question?

B) The Einstein, Podolsky, Rosen (EPR) Paradox [1] has been the subject of many discussions on quantum interpretations. In 1985, Alain Aspect et. al. measured the polarizations of two correlated photons at various rotation positions of his detectors [3, 4, 5]. Their *coincidence count* violated Bell's Inequality and experimentally verified that the nonlocal nature of quantum theory is real. The main question here is: How can two particles apparently "instantaneously" communicate with each other while being physically separated by an arbitrary distance?

C) In one type of quantum eraser experiment, polarization-correlated photon pairs are emitted in opposite directions [11]. (Refer to Figure 3 later in Section 6.) They are allowed to recombine and interfere at a 50% beam splitter, with detectors at each output port. The apparatus is set up and phase controlled such that both photons of each pair are detected by the same detector, as predicted by QM. Now, a half-wave plate is inserted in one of the paths so as to rotate the polarization by 90° with respect to the other path. This removes the guarantee of interference at the beam splitter, and it becomes possible to get one "click" at each detector from a pair of correlated photons. Building on this further, inserting linear polarizers at 45° (with respect to each photon path) directly in front of each detector returns the original interference results. It is amazing that we no longer see the effect of the half-wave plate. We have essentially erased its effect. The deepest question raised by this experiment seems to be: Can we affect the nature of the photons in the past by manipulating the present?

Current Interpretations

The myriad of interpretations of quantum theory are all attempts to explain the meaning of reality consistent with what we know about how the quantum world behaves. All popular interpretations are consistent with quantum theory and experimental results, and acceptance of any given interpretation implies acceptance of certain consequences regarding objective reality. We will mention a few of the more popular ones, but others can be found in the references.

1. The Copenhagen Interpretation denies any deep physical reality. Since elementary quanta cannot have simultaneous values for non-commuting observables, reality itself cannot be defined until a measurement is made.
2. The Many Worlds Interpretation accepts the existence of an infinite number of universes (complete with physical energy like our own). Several new ones are created every time a quantum system is forced into one of its eigenstates after previously existing in a coherent superposition of any basis states.
3. The Many Histories Interpretation establishes that trajectories taken by elementary quanta are obtained by "summing over the possible histories." For example, in the quantum erasure experiment, the linear polarizers have changed the set of possible histories of the photons' behaviors, thus affecting the possible outcomes of the experiment.
4. The Transactional Interpretation supports the notion that the emission and absorption of energy quanta is an indivisible, fundamental event. The nature of these

events, such as when and where, are determined a priori outside of time and before the transaction takes place. Note that under this interpretation, a photon absorbed by your skin from a star 1 million light years away was a predetermined event, 1 million years ago.

5. The Neorealist Interpretation maintains that the world is made up of ordinary objects as we are used to but permits that some of these objects move faster than the speed of light. Consequences of this interpretation include the possibility of reverse causality.
6. The No Collapse Interpretation maintains that the wavefunction describing a quantum system never collapses. We only observe and "think" that it has collapsed, while all the other non-measured states of the wavefunction are forever inaccessible. After the measurement, the continuing wavefunction no longer describes the probability of what can be measured. As an aside comment: the mathematics used in this interpretation begins with wavefunctions with several vector spaces involved, such as a measuring apparatus. Classical correlations emerge when one only studies the states of the apparatus. A major problem is that the mathematics needs to assume "collapse" theory to even begin writing down an initial wavefunction. Otherwise, each quantum has an incredibly complex past wave-function whose components in each space are likely to possess orthogonalities, destroying quantum inter-

ference. In short, this interpretation contains initial conditions which seem inconsistent with its conclusions.

All of these interpretations require belief in realities beyond the scope of scientific experiment. This is the price paid for claiming that the formal theory of QM describes all physical processes. The ProWave Interpretation pushes quantum weirdness back into the realm of physics, and therefore does not force us to postulate and philosophize about inaccessible realities.

The ProWave Interpretation

ProWave makes no assumption of localization of the photons before measurement. In fact, it rejects the common notion of wave-particle duality. Recent teaching by laser physicist W.E. Lamb Jr. supports this line of thinking [13]. Maintaining any sort of particle nature of elementary quanta is what has led us into trouble, philosophically. Let's start with a list of the assumptions made in the ProWave Interpretation:

1. Elementary quanta of matter and energy exist as their wavelike behavior suggests (wave packets), always.
2. Their time evolution is described by the Schrödinger equation (or better yet, by the Heisenberg equation of motion for the density operator).

3. Energy transfer, in quantum amounts, takes place locally. Thus, when a photon is absorbed and measured, its energy is <u>transferred</u> at only one point in space (Basically, this is only a defining property of "quantum").

Assumption (2) deals with propagation, which is a nonlocal, wavelike phenomenon, and Assumption (3) deals with the creation and destruction of quanta, which is a local phenomenon. These assumptions eliminate the ambiguity of what the electrons and photons do while we don't measure them. They are quantum waves: Electrons orbiting atoms are standing waves of matter-energy, photons are wavepackets of electromagnetic energy, etc. Also, we accept that a photon passes through both slits if both are probable, and it was not already absorbed by the wall. The same goes for electrons. In this way, we can reinstate physical reality which is dismantled in any wave-particle duality interpretation. The physical reality is simply that described by the wavefunctions and density matrices of a system, no matter how quantum entangled the states may be. It may be difficult to imagine the states in classical terms, but they are states nonetheless in which these very simple elementary quanta can exist. And they propagate according to the laws described by the mathematics of quantum mechanics.

As the quanta propagate and interact with the macroworld, two separate types of interactions occur. The first is defined as a partial interaction: This interaction reorganizes or redirects the wavefunction designated by a unitary transformation matrix. Examples of such are beamsplitters and magnetic fields. The other type of interaction is defined as a complete interaction: This is designated by the destruction (and creation) of a quantum of energy, for example a bound electron absorbing a photon.

Before applying ProWave to the experiments described earlier, here it's quickly shown that ProWave can add insight into how a cloud chamber can measure the particle-like nature of matter waves. As, say, an electron traverses its ``path" in the cloud material, it is constantly being forced into localized positions by partial interactions with the material. Thus, the matter wave is being reorganized constantly and not really allowed to diffract much before collapsing repeatedly. The result is a clearly drawn path that was previously believed that only a particle could make.

The ultimate challenge for ProWave is to explain how a spread-out wave can collapse and deliver its energy locally upon absorption. It is helpful to envision wave evolution analogous to blowing up a balloon. Upon measurement, that balloon pops. The collapse (also pertains to reconfiguring the wave's energy) is probabilistic, like not knowing where a balloon will pop first. But once a measurement has been taken (or perhaps the quantum is destroyed) the wave collapses everywhere nonlocally and passes its energy to the absorber as one quanta. Likewise, when a balloon pops, the entire surface of the balloon is quickly affected by the loss of tension in the rubber, however time elapses before the surface collapses. This collapse (for the quantum) need not be instantaneous but must be faster-than-light to ensure that nonlocality still applies. The phenomena of energy absorption and reconfiguration (quantum state changes) are inherently quantum uncertain events and cannot be pinned to "instantaneous"- only for practical purposes do we assume so. This is not a problem physically because the nature of this collapse is very poorly understood. In fact, I am suggesting the possibility of a more general theory that reduces to QM when the time of the collapse is considered small, in the same way that QM reduces to classical mechanics under certain conditions. This is typically the way physics advances; I don't see this potential for the other, far less intuitive, interpretations.

Explanation of the Experiments

ProWave offers new insights into the experiments involving quantum phenomena, and how nature can produce the results that it does. It is worthwhile to describe how the experimental results mentioned above can make sense given the ProWave view of reality. A deeper, more mathematical description of the experiments is found in the Appendix which follows with ProWave providing a consistent, realistic explanation.

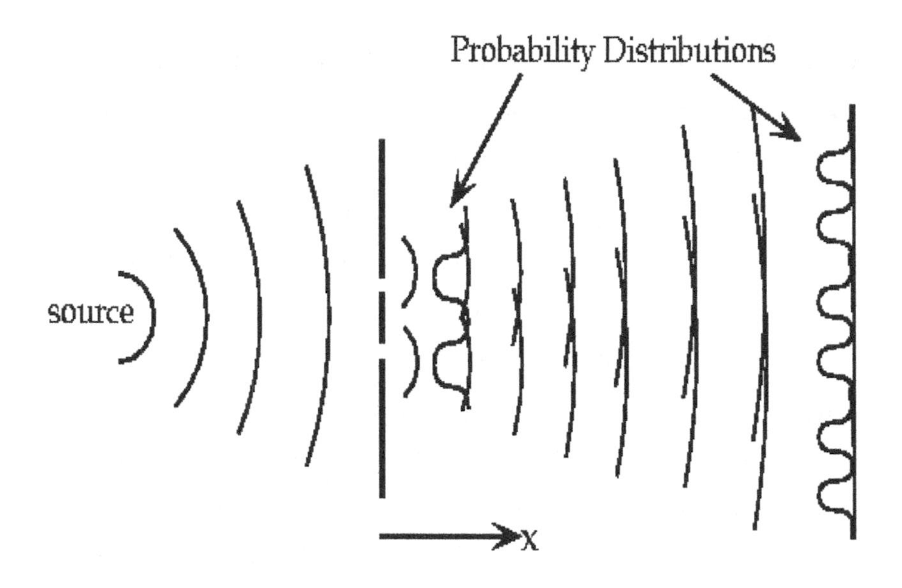

Figure 1: A typical 2 slit experiment showing probability distributions at different distances from the slits.

In the two-slit experiment (Fig. 1), a photon enters the two slits as a single, expanding wave. At the wall, its instantaneous probability of impacting on the surface of the wall is quite high. In the event that it makes it through the slits, the wave at the walls disappears and all the energy is collected and passes through the slits. This is governed by the quantum mechanical probability of absorption of the wall. As the wave now interferes with itself upon passing through the slits and propagating some distance to allow overlap, the energy of the quantum is spread out according to its probability distribution function (Ψ^2). Absorption of the photon, now, is a local process governed again by probability. Once the probability distribution collapses, the "balloon pops" everywhere and the quanta of energy localizes at the point of transfer. This localization need not occur instantaneously, and it makes sense to conceive of it occurring on the timescale of energy absorption. As of today, this notion does not violate any laws of physics because there currently is no description of how a photon's probability distribution collapses upon absorption.

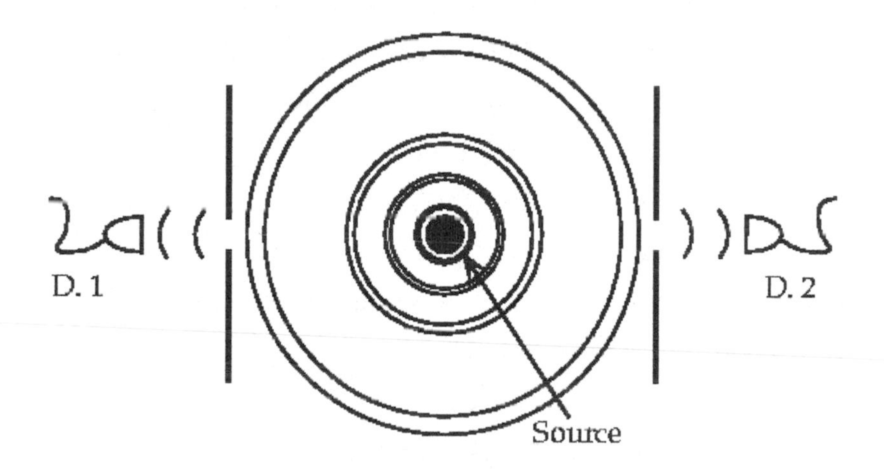

Figure 2: An experiment with EPR correlated particles.

I think the key to understanding any of the EPR correlations (Fig. 2) is to forego the idea that two distinct particles are emitted from some source in opposite directions. Quantum mechanics treats them as indistinguishable and as both being emitted in both directions, so let's treat them that way physically. The sources typically do not prefer a direction upon emission, making the correlated photons themselves spherical waves propagating, overlapping each other (one field). Any effect on one wave has a direct effect on the other wave as well. Thus, say, if a particle is measured at one location (a photon was absorbed locally), then its partner snaps into a state opposite the source and resumes propagation. Analogously, imagine two balloons expanding together with the same radius. If one balloon pops, that causes the other to reconfigure as well. The reconfiguring of the second photon could be conceived of to have occurred instantaneously, but the final act of localization for energy transfer need not be. Again, this is consistent with what is known from QM about spin, polarization, and momentum correlations.

The quantum erasure experiment (Fig. 3) described in Section III is nothing other than a "directional interference" effect. No future activities or consciousness effects are needed to explain this experiment. The correlated pair is emitted as two quanta overlapping each other. They interfere at the splitter and the effects are detected. When one path undergoes polarization rotation, we no longer observe the interference of the two paths. This is no surprise because one path has polarization x and one has polarization y, and these are orthogonal. However, both x and y polarizations have a component of polarization in the z direction. These z-components do interfere at the beam splitter and the effect can be observed by putting in a linear polarizer oriented in the z direction directly in front of each detector. Note that the z-polarizer effectively filters out the orthogonal ($\bar{z}$) interference which can cancel the z-component interference. In essence, all of these interference effects are present until the complete interaction. The type of interference that is manifested and becomes visible is a function of how the detectors are set up.

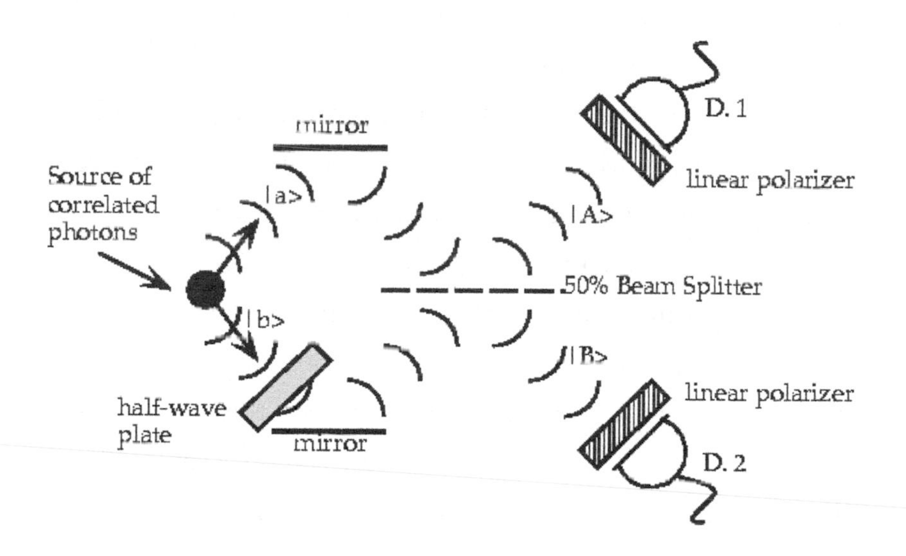

Figure 3: The quantum eraser experiment.

Summary

Photons, electrons, and all other elementary building blocks of our physical world represent simplicity in nature and existed long before complex structures like humans existed. Reality is quite real for all forms of matter, although admittedly, it is not always intuitive to rationalize the nature of this simplicity. What is meant by "simplicity" here is that just because we need strange quantum mechanical language to describe these systems, quantum mechanics can describe quite accurately their behavior, for example, "spin up" and "spin down" states. In contrast, to accurately describe a more macro structure such as a group of several molecules, a hopelessly more complex calculation is needed. Thus, the quantum world represents simplicity, not complexity.

Whether in the form of leptons or photons, individual quanta of energy are passive creatures and propagate as waves, interfering as expected. These effects are observed only by carefully controlling the "directions" (e.g., spin projection, and polarization) in which we choose to observe the waves. The mystery of quantum behavior occurs in the waves' ability to collapse into eigenstates upon absorption or possibly a partial interaction with other matter. There is no wave-particle duality. There are no point particles, only localized energy transfers. The ProWave Interpretation of quantum mechanics is the alternative to the myriad of unpalatable existing interpretations. The building blocks of matter and radiation exist in only one world, and they have only one history.

Appendix

This section contains mathematical descriptions of the experiments discussed in the text. The ProWave Interpretation is consistently applied here and offers an intuitive way to digest the equations.

A) To better describe the double slit experiment, we will need to consider the quantum state of the slits (measure which slit through which the photons pass). It is generally understood that when the particular slit of passage is known regarding each photon, the interference pattern disappears. In the right operating regime, with the screen up close, there is no interference, with the screen far away, there is interference, as seen in the two probability distributions in Fig. 1. Consider the state of a perfect transmission:

$$|\Psi_1\rangle = \frac{1}{\sqrt{2}}(|1\rangle + |2\rangle)$$

where state vectors $|1\rangle$ and $|2\rangle$ refer to the slit number. Consider now the effect of the wall containing the slits, which tells us a number based on the slit passed by the photon. The state of this system is written as

$$|\Psi_2\rangle = \frac{1}{\sqrt{2}}(|1\rangle |1_A\rangle + |2\rangle |2_A\rangle)$$

where A = Apparatus. Thus, to describe a partial effect of the apparatus, a density matrix is needed to describe the system. There is uncertainty in the wavefunction itself, as the screen is positioned at various distances from the slits.

$$\rho = \sum_i P_i(x) |\Psi_i\rangle \langle\Psi_i| = \begin{bmatrix} \frac{1}{2} & c(x) \\ c(x) & \frac{1}{2} \end{bmatrix}$$

where $c(x)$ represents the degree of coherence between the [slit] states for the wave. Using the language of QM, but with simplified equations, let our observable = B = the "interference operator." Let a = "propagation not guided by interference" and b = "propagation guided by interference." We define B to have the following expectations:

$$\langle 1 \,|\, B \,|\, 1 \rangle = \langle 2 \,|\, B \,|\, 2 \rangle \equiv a$$

$$\langle \Psi_1 \,|\, B \,|\, \Psi_1 \rangle = \frac{1}{2}(\langle 1| + \langle 2|)B(|1\rangle + |2\rangle) \equiv b$$

The matrix

$$B = \begin{bmatrix} a & b-a \\ b-a & a \end{bmatrix}$$

has these properties. Now, using the state $\rho(x)$ we take a measurement:

$$\langle B \rangle = Tr[\rho B] = (1 - 2c)a + (2c)b$$

which provides us the expectation by which many quanta produce fringes. The coefficient of "b" indicates the relative amount of interference present at the screen.

Consider the two extreme cases:

1. Near Field: Diffraction of the waves has not caused significant overlap. $c(x) \sim 0$, and there is no interference.
2. Far Field: Waves from the two slits (but the same quantum) have a large amount of overlap at the screen. $c(x) \sim \frac{1}{2}$ gives rise to the quantum coherence and so interference fringes emerge.

Interpretation: It is the present state of the system which describes the possible outcomes, not our conscious "knowledge" of anything regarding, say, the "path taken by the particle." In this particular set-up, any photon not hitting the wall containing the slits passes through both slits. Note, a partial interaction between matter and a quantum of energy puts that quantum into a state we cannot adequately conceptualize (i.e. a reconfiguration of the energy in the wave packet), but we can only intelligently discuss probabilities of likely outcomes upon a complete interaction with matter in the case of photons.

B) Many EPR experiments involve the generation of correlated pairs of photons in an entangled state. Without loss of generality, consider as an example, the singlet state

$$|\Psi_0\rangle = \frac{1}{\sqrt{2}}(|x_L\rangle\,|x_R\rangle + |y_L\rangle\,|y_R\rangle)$$

Where *x,y* refer to orthogonal polarization states, and *L,R* label the left and right paths. The results of these experiments pioneered by Alain Aspect in 1982 confirm nonlocal effects. All the mathematics confirms this as well, but it has also been proven that these effects cannot be used to change the statistical averages of observables over distances faster than light could propagate a signal [7]. Any observable, *B*, which acts on one of the two sides treats this state as if each particle were in a mixture of states and not quantum correlated. Only after bringing together the data from both detectors can the patterns of quantum interference be observed. Neither side will exhibit quantum interference on its own, for example, measure any observable *B* on the left side:

$$\langle B_L \rangle = \langle \Psi_0 \,|\, B_L \,|\, \Psi_0 \rangle = \frac{1}{2}(\langle x_L \,|\, B_L \,|\, x_L \rangle + \langle y_L \,|\, B_L \,|\, y_L \rangle)$$

The interference terms with inner products involving both x and y disappear from the orthogonality of $\langle x_R | y_R \rangle = 0$. As Fig. 2 suggests, these quanta are waves propagating and overlapping in space. Thus, they are both present at both detectors. The measurement of one quantum (by probabilistic means) forces the other to reconfigure opposite to the first one. When the first quantum undergoes annihilation and transfers its energy locally, the other quantum reconfigures as well, governed by the quantum correlation (which includes conservation laws). Based on the equation above, by measuring the polarization of the photon on the left, we can determine what the polarization of the photon on the right will be if measured in the x-y basis, but we cannot use this knowledge to send information faster-than-light.

C) The Quantum Eraser Experiment can be understood in terms of following the state functions of the correlated particles. First, let's understand how the interference is observed without the polarization rotator. With the rotator removed, the state can be expressed as

$$|\Psi_0\rangle = \frac{1}{\sqrt{2}}(|a_1\rangle\,|b_2\rangle + |b_1\rangle\,|a_2\rangle)$$

where *a,b* refer to the two different paths described in (Fig. 3), and 1,2 index the quanta. The beam splitter performs a unitary transformation into the *A,B* basis according to

$$U = \frac{1}{\sqrt{2}}\begin{bmatrix} 1 & 1 \\ 1 & -1 \end{bmatrix}$$

where *U* acts on each *a,b* state (and each photon) separately giving

$$U|a\rangle = \frac{1}{\sqrt{2}}(|A\rangle + |B\rangle)$$
$$U|b\rangle = \frac{1}{\sqrt{2}}(|A\rangle - |B\rangle).$$

Thus, upon leaving the beam splitter, the 2-photon state is

$$|\Psi\rangle = U_1U_2|\Psi_0\rangle = \frac{1}{\sqrt{2}}(|A_1\rangle|A_2\rangle - |B_1\rangle|B_2\rangle)$$

which indicates that the only two outcomes of a measurement are both quanta appearing at the first detector, or both at the second detector. ProWave asserts that at this point, both photons are still present at both paths A and B, like the wave function suggests, until the energy is actually transferred to the detector. This is because there is quantum coherence between these states.

Now, if we insert the polarization rotator in the "b" path, the system acquires an additional degree of freedom and is described by

$$|\Psi_1\rangle = \frac{1}{\sqrt{2}}(|aP_+\rangle_1|bP_-\rangle_2 + |bP_-\rangle_1|aP_+\rangle_2).$$

Where P_+ and P_- represent orthogonal polarization states. Next, we use this state to propagate through the beam splitter, obtaining

$$
\begin{aligned}
|\Psi'\rangle &= U_1 U_2 |\Psi_1\rangle \\
|\Psi'\rangle &= \frac{1}{2\sqrt{2}}(|AP_+\rangle_1 |AP_-\rangle_2 - |AP_+\rangle_1 |BP_-\rangle_2 \\
&\quad + |BP_+\rangle_1 |AP_-\rangle_2 - |BP_+\rangle_1 |BP_-\rangle_2 \\
&\quad + |AP_-\rangle_1 |AP_+\rangle_2 + |AP_-\rangle_1 |BP_+\rangle_2 \\
&\quad - |BP_-\rangle_1 |AP_+\rangle_2 - |BP_-\rangle_1 |BP_+\rangle_2).
\end{aligned}
$$

With all these terms present, it is no surprise that this state generates all combinations of detector clicks (with repeated events, of course). Each term in this state has equal likeliness of emerging upon detection.

With this same state, $|\Psi'\rangle$, consider putting in the two linear polarizers in front of the detectors, oriented at 45° to both beams for simplicity. These polarizers let pass the polarization state $|P_0\rangle$ and absorb the orthogonal state $|\overline{P_0}\rangle$ and are related to the original polarization basis by

$$
\begin{aligned}
|P_+\rangle &= \frac{1}{\sqrt{2}}(|P_0\rangle + |\bar{P}_0\rangle) \\
|P_-\rangle &= \frac{1}{\sqrt{2}}(|P_0\rangle - |\bar{P}_0\rangle).
\end{aligned}
$$

To find out what effect these have on the state of the system, project onto the polarizers' basis, and drop all terms that don't lead to coincidence counts (2 photon counts). The new wavefunction now has only P_0 terms and is not normalized because of the dropped terms.

$$|\Psi''\rangle = \frac{1}{2\sqrt{2}}(|A_1\rangle |A_2\rangle - |B_1\rangle |B_2\rangle).$$

Thus, we have recovered the same interference effect as before by treating the wavefunction as the actual energy carried by the photons right up to the point of their demise in the detector. The state of the photons is the reality we seek, and it describes all interactions and interferences we observe with a causal relationship. The interference along all directions was always present, up till the final interaction which destroyed the photons.

Acknowledgments

I thank all the members of our "quantum lunch" group: Joe Provenzano, Erann Gat, Joe Kahr, and Rich Doyle for our stimulating discussions.

Dan Provenzano
Thu Jan 15 19:33:34 PST 1998

References

1. Einstein, B. Podolsky, and N. Rosen, Phys, **47**, 777 (1935).
2. J. S. Bell, Physics (N.Y), 1, 195 (1965).
3. Aspect, P. Grangier, and G. Roger, Phys. Rev. Lett. **47**, 460 (1981).

4. Aspect, P. Grangier, and G. Roger, Phys. Rev. Lett. **49**, 91 (1982).
5. Aspect, P. Grangier, and G. Roger, Phys. Rev. Lett. **49**, 1804 (1982).
6. J. Horgan, Sci. Amer. July, 1992.
7. G. C. Ghirardi, A. Rimini, T. Weber, Lett. Nuovo Cimento, **27**, 293 (1980).
8. D. Mermin, Am. J. Phys. **49**, 940 (1981).
9. D. Mermin, Phys. Today, Apr. 1985.
10. J. Pykacz, Phys. Lett. A. **171**, 141 (1992).
11. P.G. Kwiat, A.M. Steinberg, and R.Y. Chiao, Phys. Rev. A. **45**, 7729 (1992).
12. N.J. Cerf and C. Adami, "Quantum Information Theory of Entanglement," Phys. Comp. 96. (1996).
13. W.E. Lamb, Jr., Appl. Phys. B. **60**, 77 (1995).

Made in the USA
Monee, IL
26 September 2022

14396840R00079